Springer Texts in Social Sciences

This textbook series delivers high-quality instructional content for graduates and advanced graduates in the social sciences. It comprises self-contained edited or authored books with comprehensive international coverage that are suitable for class as well as for individual self-study and professional development. The series covers core concepts, key methodological approaches, and important issues within key social science disciplines and also across these disciplines. All texts are authored by established experts in their fields and offer a solid methodological background, accompanied by pedagogical materials to serve students, such as practical examples, exercises, case studies etc. Textbooks published in this series are aimed at graduate and advanced graduate students, but are also valuable to early career and established researchers as important resources for their education, knowledge and teaching.

The books in this series may come under, but are not limited to, these fields:

- Sociology
- Anthropology
- Population studies
- Migration studies
- Quality of life and wellbeing research

Dimitri Mortelmans

Doing Qualitative Data Analysis with NVivo

Springer

Dimitri Mortelmans
Faculty of Social Sciences
University of Antwerp
Antwerp, Belgium

ISSN 2730-6135 ISSN 2730-6143 (electronic)
Springer Texts in Social Sciences
ISBN 978-3-031-66013-9 ISBN 978-3-031-66014-6 (eBook)
https://doi.org/10.1007/978-3-031-66014-6

This Springer imprint is published by the registered company Springer Nature Switzerland AG
The registered company address is: Gewerbestrasse 11, 6330 Cham, Switzerland

Preface

Qualitative research has known turbulent times. The War of the Methods questioning the scientific status of qualitative research lies behind us. The suspicion of digitalisation and colonisation by software programs has silenced. The rise of Mixed Methods was embraced and positioned as a middle way between the qualitative and quantitative approaches of scientific inquiry. And through all these academic debates, Computer-Assisted Qualitative Data AnalysiS (or CAQDAS) developed (Paulus et al., 2014), almost unnoticed. Starting with a wide range of competing programs, the market compacted to a limited number of players. One of the leading software tools is the topic of this comprehensive guide: NVivo. NVivo started as a program designed to code (textual) data and retrieve coded references. Throughout the years, it expanded to a leading software tool to work with multiple data sources, including audio, video and surveys; designed to execute the most complex queries on coded data and to visualise the results in multiple ways. It helps researchers to unlock the potential of their qualitative data, offering rich insights and a deeper understanding of the data they have collected.

The primary aim of this book is to bridge the gap between the daily practice of a qualitative researcher and the software they use. The book does not have the purpose to be a mere software manual or button-course. Off course, when writing about software, buttons, menus and screenshots are indispensable. But all buttons have already been described in the NVivo help pages or on YouTube in the introductory videos. This book grew out of many years of giving Ph.D. workshops on NVivo. Throughout the years, we learnt how qualitative researchers experienced their first encounter with NVivo. We learnt about their struggles with a new piece of software. From that feedback, we learnt to present NVivo from a researcher's perspective. And that is the major aim behind this book: to show you NVivo from the needs of the qualitative researcher, whether experienced or just starting. At the same time, we try to take the wide diversity of qualitative research into account.

Coming myself from a Grounded Theory tradition, our first (Dutch) book on NVivo was exclusively oriented to using the software to perform a Grounded Theory analysis (Mortelmans, 2011, 2017). But as NVivo developed into an all-encompassing software program, our courses were also followed by Ph.D.'s that needed more than just an introduction to Grounded Theory with NVivo. As a consequence, the narrow focus disappeared from this book and a wide range of data sources, approaches and techniques are discussed throughout the book. We start from setting up a project and importing data, to coding, querying and visualising data. We also explore more advanced features, such as classifications, and framework matrices, literature reviews and autocoding enabling a wide range of qualitative researchers to conduct complex analyses and extract meaningful insights from their data with their analytical approach.

This book is not a methods book. Some excellent introductions on NVivo (e.g. Jackson & Bazeley, 2019) combine an extensive introduction to qualitative research methodology and its application in NVivo. As this sacrifies room for more detailed insights in NVivo, we kept the methods background to a minimum. This implies that this book is not to be used in a methods course as a handbook but should help you in discovering NVivo's core tools and hidden gems and gain efficiency in using them in your daily practice.

This book is written for both novice and experienced researchers. For beginners, it provides a step-by-step guide to using the three basic skills: project management, coding data and retrieving coded data with Queries. For experienced researchers, we dive into advanced techniques like classifications or artificial intelligence in CAQDAS, and we show the latest features of the program. As such, both the beginning and the experienced users will be able to leverage the full potential of the software in their research no matter what stage of experience you have reached.

The book is structured into three parts. First, we start with a more theoretical background on qualitative research and the place of software in qualitative inquiry. Next, we explore the basic components of the program: setting up a project, importing and managing data, coding, analysing data, and presenting findings. Last, we have a set of more topical chapters where we apply the tools to specific situations like focus groups, mixed method research or literature reviews.

We would like to express our gratitude to all those who have contributed to the creation of this book. Special thanks to the team at Lumivero for developing NVivo and continuously improving it, making qualitative research more accessible and efficient. And even though we do not know them personally, thanks are also warranted for Tom and Lynn Richards for starting this program so many years ago (under the name of NUD*IST). For those interested in the origin of NVivo, we highly recommend Tom's article on the intellectual history of the program (Richards, 2002). In the journey of writing this book, we have been fortunate to be supported by the insights of many of our colleagues and collaborators. Among them, one colleague and friend stands out for his exceptional contribution: Olivier Chandesais. Olivier's thorough reading and detailed commentary on each chapter of this manuscript have been invaluable. As no one we have ever seen, his eye for detail and his year-long expertise with qualitative data analysis in teaching has not only enhanced the accuracy of this book but also enriched its content. It was with much curiosity that we looked forward to his next revision and corrections whenever he finished another chapter. Last, we want to thank all the Ph.D. students that have followed our workshops the past decades. Through their struggles with the software, they gave invaluable feedback and suggestions, which have greatly transformed our original software focus into a researcher focus. The unintended insights we got into the minds of starting qualitative researchers were invaluable to our own development as teacher and writer of this book.

Before ending this preface, we want to inform you about the graphical conventions we used when writing this book. In a book, one needs to convert the visual information on the screen of the reader into a written form. Off course, we show screenshots with explanations whenever necessary but still a verbal translation of visual information is unavoidable in this type of book. First, whenever we refer to a button or window, we will use italics for the name of the item and refer to add the name of the item (e.g. the *OK* button, the *Welcome* screen). Names of menus will be put in bold (e.g. the **File** menu). Often, you need to open levels in a menu or a folder structure. The different levels will be separated by an arrow sign: >. The different levels will be put in bold. For menu's, this will look like: **File > Project Information > Open Project Event Log**. For the Navigation View (the blue rectangular area at the left of your program screen), this will look like: *Navigation View* > **Data > Files**. When options are mentioned, they will be written in Italic. When you need to click a check-box, we will use this symbol: ☑. Even though most options are available through menus, working with the *Context menu* is often more efficient. When you right click in your program, a menu will appear next to your pointer offering options the programmers judged useful when being in that part of the program. We refer to a Right Mouse Click with RMC. Also the place on your screen to click will be indicated. So, referring to an option in the Context Menu will be done as follows: **RMC** (*above the main folder*) > **New Folder**.

Throughout the book, we will use the sample projects embedded in the software. As such, all readers of the book can reproduce the examples given in this book. As NVivo regularly gets updates, it is possible that changes in the data of these sample projects produce slightly different results than we show in the screenshots. We hope that these changes will still enable you to follow the examples even though our screenshots might differ from what you see on your screen. Also software changes through updates can create such differences although we are confident that the main structure and way of working with NVivo remains stable across updates and even across versions in the future. As the field of qualitative research and the NVivo software are constantly evolving, feedback and suggestions from you as reader are most welcome. Your feedback will help improve future editions of this book and ensure that it remains a valuable resource for qualitative researchers.

Thank you for choosing this book. We hope it will serve as a valuable guide in your journey of qualitative research with NVivo.

Antwerp, Belgium
January 2024

Dimitri Mortelmans

References

Jackson, K., & Bazeley, P. (2019). *Qualitative data analysis with NVivo.* Sage.

Mortelmans, D. (2011). *Kwalitatieve analyse met Nvivo.* Acco.

Mortelmans, D. (2017). *Kwalitatieve analyse met Nvivo.* Acco.

Paulus, T. M., Lester, J. N., & Dempster, P. G. (2014). *Digital tools for qualitative research.* Sage.

Richards, T. (2002). An intellectual history of NUD*IST and NVivo. *International Journal of Social Research Methodology*, 5(3), 199–214. https://doi.org/10.1080/13645570210146267

Acknowledgements

I want to thank all the students who have followed my courses in Qualitative Data Analysis and NVivo for their valuable feedback. Their comments made this book more oriented on what researchers want to learn when working with NVivo in their qualitative analyses.

I also explicitly thank Olivier Chandesais for the years of collaboration in all my courses, the hundreds of tips I got, and the insights I acquired from our long talks about methodology and research in the social sciences.

Parts of Chaps. 1 and 2 are reproduced from *Kwalitatieve Analyse met NVivo* by Dimitri Mortelmans with permission from Acco.

Contents

CHAPTER 1

A Guided Tour in Qualitative Research

> **Key messages in this chapter**
>
> - Qualitative research methodology is defined through its research components.
> - The place of software in qualitative data analysis is accepted but also contested by some scholars.

What is Qualitative Research?

Delineating the boundaries of qualitative research within a single definition is not straightforward. Intuitively, many perceive qualitative research as "something that does not involve numbers" or "something that includes open-ended interviews". A potential start is the definition of Denzin and Lincoln from the first edition of their handbook:

> Qualitative research is a situated activity that locates the observer in the world. Qualitative research consists of interpretive, material practices that make the world visible. These practices transform the world. They turn the world into a series of representations, including field notes, interviews, conversations, photographs, recordings, and memos to the self. At this level, qualitative research involves an interpretive, naturalistic approach to the world. This means that qualitative researchers study things in their natural settings, attempting to make sense of or interpret phenomena in terms of the meanings people bring to them. (Denzin & Lincoln, 2017, p. 10)

D. Mortelmans, *Doing Qualitative Data Analysis with NVivo*, Springer Texts in Social Sciences, https://doi.org/10.1007/978-3-031-66014-6_1

Central to this definition is the researcher's perspective on the world. Furthermore, the natural, everyday environment is pivotal in the research, and a researcher who seeks to understand processes of meaning-making. There are many more potential definitions of qualitative research, but often elements from the aforementioned definition are repeated, omitted, or supplemented. Within the scope of this book, it is not our aim to attempt to provide a definitive definition of what qualitative research is or should be. Hence, we align with a group of authors (e.g. Bryman, 2012; Shank, 2006; Snape & Spencer, 2003) who attempt to define qualitative research by listing characteristics that are often present, but not necessarily found in all qualitative research. The distinct nature of qualitative research can be found in the domain of the research questions, the employed research design, the methods of data collection, the analytic approach, and the output that the research ultimately yields.

The different components in Table 1.1 together provide an insight into the essence of qualitative research. However, it is crucial to recognize that these characteristics are largely indicative. The recurrent mention of 'flexibility' highlights the relative nature of such lists. Simultaneously, it becomes evident that qualitative research is a large 'tent' (Shank, 2006) or 'umbrella' (LaMarre & Chamberlain, 2022) under which many approaches can be sheltered.

Table 1.1 Overview of the core characteristics of qualitative research

1. *Research questions and objectives*
    - Questions address complex themes or social processes
    - The everyday reality of the subjects is central
2. *Research design*
    - The design is flexible
    - The design focuses on studies in a natural setting
    - The design aims for a "holistic" understanding of the context
3. *Data collection methods*
    - A wide range of data collection methods is available
    - Multiple methods are often employed in a single study
    - The use of methods is flexible
    - Data collection often implies intense and/or prolonged engagement with the field
4. *Analysis*
    - The analysis is primarily text-based rather than numerical
    - The goal of analysis is to uncover meaning
    - Processes are central in the analysis
    - The aim of the analysis is to understand in-depth rather than to provide representative descriptions
5. *Reporting*
    - Subjects are involved in (the review of) the results
    - Reporting attempts to represent the context of the whole
    - The influence of the researcher on their research design is explicitly considered

Research questions

In qualitative research, the focus is not on the researcher but on the researched. It is not the all-knowing scientist who is central, but the everyday living environment of the individuals or groups under study. This is the classic distinction made in anthropology between the deductive and inductive perspectives in research. A *deductive* perspective involves a researcher who, guided by literature study and previous results, approaches the subject with a predetermined framework (such as a closed questionnaire). In contrast, the *inductive* approach of qualitative research seeks these frameworks among the subjects themselves (Silverman, 2018). When a researcher adopts this approach, different research questions naturally come into focus. Instead of quantifying frequency, the researcher seeks to understand how social meaning is constructed, how social processes unfold.

The research design

When seeking to understand the meaning-making of individuals, we must be acutely aware that although humans are composed of atoms, they certainly do not react like atoms. Human nature is inherently unpredictable. However, this does not imply that the study of social interaction and processes is an impossible task. Like Miles et al. (2018), we consider ourselves "transcendental realists". This means we believe that social phenomena exist not only in our minds but also outside them, and that scientists are capable of discovering lawful and reasonably stable relationships in that objective world. These lawful and stable foundations of social life enable the development of concepts and theories that provide insight into underlying processes in social reality. This perspective is eminently suitable from both a quantitative and qualitative research viewpoint. As a qualitative researcher, you also aim to identify regularities with the goal of theory formation. This focus is much closer to the subjects themselves compared to quantitative research. Social processes and meaning-making are examined from the perspective of the individuals' own lived experiences, including the rich context of their lives. This context is not "preformed" as in quantitative research, where questions are the same for everyone, and the context - what is not asked in the standard questionnaire - is assumed to be constant. In qualitative research, there is a continuous interaction between your study and its context, where sometimes context becomes research and research becomes context. Unexpected occurrences are common during qualitative studies. To understand how people give meaning to their surroundings, and how the environment shapes this meaning-making, it is necessary to employ a research design that is as open as possible. A research design is the way a researcher plans and organizes a study in advance. It is essentially their "work plan". Such an approach needs to be '"holistic", meaning that the research design should be capable of approaching the research topic in a systematic, comprehensive, and integrated manner.

Therefore, qualitative researchers prefer not to fix everything in advance too rigidly. They review literature to understand the state of research in their field, without going so far as to prevent their minds from entering the research field with an "open" mindset. They write a research question that serves as a compass for the study, without this question precluding all side paths and discoveries. They collect data without creating instruments beforehand that are fixed and uniform for all subjects.

The data collection method

In qualitative research, one does not confine oneself to a single method of data collection or consistently use the same method. Switching data collection methods during the course of a study, as necessitated by the research setting, is common. Moreover, employing multiple methods to gather data is more the norm than the exception. In quantitative research, one usually limits oneself to either conducting an experiment or administering a structured questionnaire. In qualitative research, however, you will likely collect statistics, conduct observations, and carry out interviews. It is often observed that in qualitative research, one method tends to be dominant. The researcher may primarily focus on participating in the field or mainly conduct in-depth interviews.

A characteristic common to many qualitative data collection methods is the prolonged and in-depth contact with the field. Gathering data about the context in which people live and the meanings they ascribe to their environment is rarely a matter of minutes. It can take days, weeks, or even years to gain a deep understanding of the subject of study. Particularly when opting for observation and participation techniques, the time for data collection increases.

The analysis

Qualitative research is sometimes narrowly defined as a method that produces results without statistical procedures or quantification (Strauss & Corbin, 1998). Many students also confuse the absence of numbers with qualitative research itself. Often, a choice is made for qualitative research in a thesis or paper because it does not involve numbers, and more importantly, no statistics. While reducing qualitative research to non-numerical research is an accurate depiction, this definition conceals as much as it clarifies. The lack of statistical analysis is a common feature across almost all qualitative variants. However, the characteristics previously mentioned are at least as defining for qualitative research as the mere absence of numerical data. If a researcher poses quantitative research questions and then uses a small sample with in-depth interviews to answer these questions, they have conducted neither quantitative nor qualitative research, even if no statistics were involved. The research may be presented as quantitative, but it is merely a weak version of it.

Conversely, qualitative researchers sometimes do use numerical data to outline the context of their research problem. Statistical data can help illuminate the contours of a particular issue, indicating what the researcher is addressing. Quantitative supplementary information can also be useful in analysing qualitative material. Nevertheless, the primary material for analysis is predominantly textual. Interviews are transcribed, and these transcripts are analysed. With observation or participation techniques, field notes are created, again a textual form of primary material. Visual material is also used to a lesser extent, focusing on content rather than appearance (as in quantitative content analysis, see Berelson, 1971; Krippendorf, 2018; Neuendorf, 2017). The analysis of this visual material is again conducted with words. Qualitative researchers read, code, and interpret their data. They create concepts and construct theories based on them. They typically work inductively, without testing pre-established hypotheses.

Since the goal of the analysis is to uncover the meaning-making of individuals and gain insights into social processes, flexibility is also necessary when analyzing qualitative data. While statistical procedures might be challenging due to their mathematical background, they offer the advantage of being unequivocal and can be quickly executed using accessible software. Today, one primarily needs to know the right menus and input screens to perform complicated statistical operations. Although NVivo will help you in performing your qualitative analysis, like all qualitative software programs, it lacks predefined paths. NVivo incorporates the flexibility of qualitative research and gives the researcher the freedom to conduct their analysis as they wish or as their data guide them but there are no pre-programmed routines that immediately provide output for the researcher to use in their report.

Reporting

Lastly, the reporting in qualitative research also differs. Although qualitative research also emphasizes presenting results in an accessible and readable manner, the style of writing differs from that of quantitative research. In qualitative research, the aim is to provide an extensive description of the collected material. Throughout the description, the material should 'speak' for itself. The goal is not to present all results as compactly as possible but to offer a rich contextual sketch of the findings.

Additionally, the respondents who provided the primary material during data collection can be involved in the reporting process. Unlike in quantitative research, where respondents often learn about the results from the researcher via the media, in qualitative research, direct feedback on the results is sought from the participants. This approach checks whether the analyses conducted by the researcher align with the respondents' own perceptions.

The (Contested) Role of Software in Qualitative Research

The emergence of illuminated manuscripts during the Middle Ages stands as a testament to the era's artistic and cultural heritage. Crafted with immense care in monasteries worldwide, these manuscripts, including gospel books, psalters, and bibles, were intricately adorned with an array of initials, marginalia, and miniature illustrations. However, the advent of the Gutenberg printing process, which enabled the mass production of books, simultaneously marked the decline of the labour-intensive tradition of manually copying manuscripts. While the efficiency and large-scale capabilities of printing technology are undeniable, they inadvertently signalled the end of the age-old art of illuminated manuscripts. A similar trajectory is observable with the introduction of Computer Assisted Qualitative Data AnalysiS (CAQDAS). This technological innovation, much like Gutenberg's printing press, brought about significant changes in the field, hinting at a transformation akin to the one experienced by the art of manuscript illumination.

"Invent the piano, and a whole host of composers will start writing a new music." (Richards, 2002, p. 203). Richards writes this quote to illustrate the huge impact (his) software had on the daily practice of qualitative research. The quote draws a parallel to the evolution in writing and book production, positioning the development of Computer Assisted Qualitative Data AnalysiS (CAQDAS) within a narrative of both progress and loss. While the adoption of software in qualitative research is now widely regarded as standard practice, it's important to acknowledge that this technological integration does more than just offer benefits. It may also have fundamentally altered the essence of qualitative analysis itself.

Several concerns have been raised by authors about the application of software in analysing qualitative data. The first issue pertains to the *legitimation of the analysis* through software use. Authors like Bong (2007) have noted that merely mentioning software like NVivo in a study's methodology does not inherently validate the analysis. As Barbour (2001) and Pratt (2009) caution, simply citing software, similar to general terms like "purposive sampling" or "Grounded Theory", does not guarantee analytical rigor. Editors increasingly encourage authors to move beyond this simplistic validation approach.

A second concern relates to the confusion between qualitative data analysis and coding. Seidel (1991) referred to this as "analytical madness", highlighting the risk that the ease of coding with software could lead to an overemphasis on coding as an end goal, rather than a means to achieving theoretical depth. This is echoed by Levin (1986), who suggests that software might transform from a support tool to a methodology in itself. Coffey et al. (1996) extend this argument, noting that the popularity of CAQDAS, predominantly developed by Grounded Theory researchers, risks pushing qualitative research towards a uniform approach. However, as Lee and Fielding

(1996) and Leech and Onwuegbuzie (2011) argue, most modern software packages offer flexibility that accommodates various analytical methods.

Thirdly, Weaver and Atkinson (1994) argue that software can *distance the researcher from his data*, a phenomenon known as "reification of researcher and data". This issue arises from coding processes that may detach researchers from the original context of their raw material (e.g. a transcript). However, NVivo addresses this by allowing broader context retrieval during coding. When reviewing coded material with queries, the researcher can use the option "spread to" to increase the original coded material to a broader context (see Section "Dissecting the Query Window" in Chap. 12).

Additionally, there are concerns about the expanding size of qualitative research projects, now that NVivo is capable of handling vast data quantities. This growth, as Jansen (2005) points out, has led to the term 'qualitative survey' being coined. However, this expansion raises fears of a shift towards quantitative analysis methods, where the focus is more on counting codes rather than in-depth interpretation (Bassett, 2004; Mason, 1996). NVivo's development over time reflects this risk, with increased capabilities for statistical analysis and mixed methods. NVivo does produce several statistics on number of codes, number of fragments coded at and the percentage of text (or visuals) being coded. As such, the software supplies abundant opportunities to fall in this trap. But the risk is bigger than the information given in the codebook. The data can be exported to Excel or SPSS leading to a further quantification of the coding work (see Section "Exporting Project Material with Text Reports (Extracts)" in Chap. 14) and NVivo also imports surveys from Qualtrics or Survey Monkey (see Section "Importing Primary Data in Your Project" in Chap. 6). In addition, quantitative techniques like cluster analysis are built into the program to help users automatically detect (statistical) associations within the raw material (see Chap. 19). This evolution is driven by a rhetoric of the programmers that NVivo increasingly will help you to detect the patterns in the data automatically instead of helping the researcher to dive into his data and support the intellectual analysis of the empirical material.

It's worth noting that most of these concerns, as highlighted by authors like Barry (1998), date back to the 1990s when CAQDAS tools were new and less developed. But even back then, the benefits of using such software, including efficiency and flexibility, were already acknowledged by authors like Tesch (1990, 1991). However, the earlier warnings still hold relevance. As NVivo offers more automated and AI-based analysis features, there's a risk that researchers might opt for quicker, less rigorous methods over more detailed manual coding processes. The rise of Generative AI and its application in qualitative data analysis (especially coding data) is reigniting the debates about the role of software in qualitative research (e.g. Davison et al., 2024).

References

Barbour, R. S. (2001). Checklists for improving rigour in qualitative research: A case of the tail wagging the dog? *BMJ, 322*(7294), 1115–1117. https://doi.org/10.1136/bmj.322.7294.1115

Barry, C. A. (1998). Choosing qualitative data analysis software: Atlas/ti and Nudist compared. *Sociological Research Online, 3*(3). http://www.socresonline.org.uk/3/3/4.html

Bassett, R. (2004). Qualitative data analysis software: Addressing the debates. *Journal of Management Systems, 16*(4), 33–39.

Berelson, B. (1971). *Content analysis in communication research*. Hafner Publishing Company.

Bong, S. A. (2007). Debunking myths in CAQDAS use and coding in qualitative data analysis. Experiences with and reflections on Grounded Theory Methodology. *Historical Social Research, 19*, 258–275.

Bryman, A. (2012). *Social research methods* (4th ed.). Oxford University Press.

Coffey, A., Holbrook, B., & Atkinson, P. (1996). Qualitative data analysis: Technologies and representations. *Sociological Research Online, 1*(1), http://www.socresonline.org.uk/1-1/4.html.

Davison, R. M., Chughtai, H., Nielsen, P., Marabelli, M., Iannacci, F., van Offenbeek, M., Tarafdar, M., Trenz, M., Techatassanasoontorn, A. A., Díaz Andrade, A., & Panteli, N. (2024). The ethics of using generative AI for qualitative data analysis. *Information Systems Journal*. https://doi.org/10.1111/isj.12504

Denzin, N. K., & Lincoln, Y. S. (Eds.). (2017). *The Sage handbook of qualitative research* (5th ed.). Sage.

Jansen, H. (2005). De kwalitatieve survey. Methodologische identiteit en systematiek van het meest eenvoudige type kwalitatief onderzoek. *Kwalon, 10*(3), 15–34.

Krippendorf, K. (2018). *Content analysis*. Sage Publications.

LaMarre, A., & Chamberlain, K. (2022). Innovating qualitative research methods: Proposals and possibilities. *Methods in Psychology, 6*. https://doi.org/10.1016/j.metip.2021.100083

Lee, R. M., & Fielding, N. (1996). Qualitative data anlaysis: Representations of a technology: A comment on Coffey, Holbrook and Atkinson. *Sociological Research Online, 1*(4). http://www.socresonline.org.uk/1/4/lf.html

Leech, N. L., & Onwuegbuzie, A. J. (2011). Beyond constant comparison qualitative data analysis: Using NVivo. *School Psychology Quarterly, 26*(1), 70–84. https://doi.org/10.1037/a0022711

Levin, R. B. (1986). Technological determinism in social data analysis. *Computers and the Social Sciences, 2*, 201–206.

Mason, J. (1996). *Qualitative researching*. Sage.

Miles, M. B., Huberman, A. M., & Saldaña, J. (2018). *Qualitative data analysis: A methods Sourcebook*. Sage

Neuendorf, K. A. (2017). *The content analysis guidebook* (2nd ed.). Sage. https://doi.org/10.4135/9781071802878

Pratt, M. G. (2009). From the editors: For the lack of a boilerplate: Tips on writing up (and reviewing) qualitative research. *Academy of Management Journal, 52*(5), 856–862. https://doi.org/10.5465/AMJ.2009.44632557

Richards, T. (2002). An intellectual history of NUD*IST and NVivo. *International Journal of Social Research Methodology, 5*(3), 199–214. https://doi.org/10.1080/13645570210146267

Seidel, J. V. (1991). Method and madness in the application of computer technology to qualitative data analysis. In N. Fielding & R. M. Lee (Eds.), *Using computers in qualitative research* (pp. 107–116). Sage.

Shank, G. D. (2006). *Qualitative reserarch. A personal skills approach.* Pearson.

Silverman, D. (2018). *Doing qualitative research.* Sage.

Snape, D., & Spencer, L. (2003). The foundations of qualitative research. In J. Ritchie & J. Lewis (Eds.), *Qualitative research practice* (pp. 1–23). Sage.

Strauss, A. L., & Corbin, J. (1998). *Basics of qualitative research: Grounded Theory procedures and techniques* (2nd ed.). Sage.

Tesch, R. (1990). Qualitative research: Analysis types and software tools. *The Falmer Press*. https://doi.org/10.4324/9781315067339

Tesch, R. (1991). Computer programs that assist in the analysis of qualitative data: An overview. *Qualitative Health Research, 1*(3), 309–325.

Weaver, A., & Atkinson, P. (1994). *Microcomputing and qualitative data analysis.* Avebury.

CHAPTER 2

A Guided Tour in Qualitative Data Analysis

Key messages in this chapter

- In qualitative research, data analysis can be done from many methodological approaches.
- Nevertheless, most approaches have the same way op analysing qualitative data.
- We illustrate one way of analysing as an example: Grounded Theory analysis.

Approaches in Qualitative Data Analysis

The analysis phase of a study is often where researchers find themselves unexpectedly challenged. While qualitative research is perceived as quick and cost-effective, especially in comparison to large-scale surveys, this is not always the case. Although data may be gathered more swiftly, the time saved is often negated by the extensive and time-consuming analytic process. This intensive engagement with the data contrasts sharply with the click-of-a-button efficiency offered by statistical software like SAS or SPSS. In qualitative data analysis (QDA), the researcher grapples with their data alone, and no computer can replace the creative phase of this analysis. Though quantitative research also demands creativity, it is largely intertwined with the computational power of the processes. Software like NVivo can ease the burden of qualitative researchers, but it will never autonomously produce results ready

D. Mortelmans, *Doing Qualitative Data Analysis with NVivo*, Springer Texts in Social Sciences, https://doi.org/10.1007/978-3-031-66014-6_2

for reporting. Consequently, delivering you with a cookbook approach to qualitative analysis is not possible. Just as one cannot encapsulate "falling in love" in rules and procedures, qualitative analysis cannot be fully captured in a definitive analytical pathway. However, this does not diminish the value of NVivo as a tool that can enhance the qualitative researcher's experience. While it doesn't provide ready-made analyses, the computational power of the software is harnessed to organize, make searchable, and visually condense qualitative material. This chapter provides an overview of the steps a qualitative researcher might take in their analysis, offering a potential pathway for researchers. This is not to suggest that it is the only or an infallibly successful approach. Every qualitative research project and set of data is unique, demanding a creative and adaptable approach to the methodologies outlined in this chapter: utilize them as needed and modify them when necessary. The flexibility of NVivo ensures that most alternative approaches can be seamlessly integrated into its usage.

Over the years, qualitative research has evolved into an amalgamation of various approaches and schools of thought. This complex evolution has led to the emergence of three distinct methodological paradigms: postpositivism, constructivism, and the paradigm encompassing critical and feminist approaches. Within the latter, various sub-streams can be identified. Each of these streams emphasizes different aspects within qualitative research and brings different focuses to the forefront. As a result, the qualitative analysis conducted by each stream can also differ.

Qualitative researchers have access to multiple sources of data. They can conduct interviews or focus groups, engage in participatory observation, gather documents, or take photographs. Theoretically, all these content streams and data forms could be placed in a matrix to derive a multitude of analytical approaches. However, this is not the case in the practical world of research. Still, it can be said that various established methods of analysis have crystallized over time. For instance, certain forms of analysis are based on specific types of data, like participatory observation (as illustrated by Spradley, 1980) or are aligned with a particular qualitative paradigm, such as phenomenology (Moustakas, 1994).

Authors like Tesch (1990) argue that the differences between these methods are not as vast as they might initially appear. Tesch categorizes the objectives of qualitative analyses into four main goals: uncovering language characteristics, discovering regularities, understanding the meaning of a text or action, and reflection. From her analysis, she proposes a continuum of methods, ranging from highly formalized (almost quantitative) to those where almost no method is defined. Beyond these extremes, she identifies ten basic principles common across most methods. Creswell (1998), in his comparison of five streams (the biographical method, phenomenology, Grounded Theory, ethnography, and case study), finds a common analytical procedure at the base of each method.

Regardless of the method or paradigm, a researcher typically starts by organizing the data (data management) and begins with reading interview transcripts or field notes. Then, the data are broken down into smaller segments, filtering out irrelevant information. This is followed by describing the data and connecting different data parts. The analysis concludes with writing up the findings. The central process in this can be metaphorically described as **dismantling** and **rebuilding**. It's like a researcher entering a jungle and encountering unknown ruins. They see the forms of buildings, some unclear or collapsed. Carefully, they remove the debris and number the stones that belong together. Then, they clean the site and begin reconstructing the buildings step by step, often returning to the leftover stones to fit them into the larger structure. The dismantling process in qualitative research is described as coding, indexing, labeling. The rebuilding phase involves linking, connecting, aggregating. The outcome of this phase are constructs, concepts, variables, themes, which later evolve into theories or narratives.

Grounded Theory as a Structured Way of QDA

In this chapter, we show you one methodological approach to CAQDAS: Grounded Theory. As we have shown above, this is only one family withing a wide range of possible analytical approaches. Grounded Theory is founded on the works of Anselm Strauss and Barry Glaser, originating from their collaborative publication "The Discovery of Grounded Theory" (Glaser & Strauss, 1967). The rationale behind this choice is its prevalence and it's historic connection to NVivo. Grounded Theory is the most expansive qualitative analytical approach (Bong, 2002). The majority of qualitative publications employ or reference Glaser and Strauss's method in data processing. The dominance of Grounded Theory is evident not only in the number of studies utilizing this method but also in the plethora of handbooks making a similar choice to this book. It is important to note, however, that the following sections are inspired by Grounded Theory but do not represent the Grounded Theory in its entirety. The divergent approaches of Glaser (1978, 1992, 2001, 2002, 2003, 2005) and Strauss and Corbin (1990, 1990, 1998a) alone illustrate that there are varying perspectives on the approach. We also opted for Grounded Theory as an example in this chapter as the predecessor of NVivo (NUD*IST) originated in this methodology. Even though NVivo today is a flexible and methodology independent program, it has its roots in the Grounded Theory approach (Richards, 2002).

Theory as the Core Component of Grounded Theory

While it is clear that Grounded Theory is not a monolithic methodology, its foundational framework shares common ground, particularly in two central elements: theory and procedures. Theoretical development has always been at the heart of the Grounded Theory approach. The very idea of the method is that the researcher develops theory based on empirical material. It shifts the focus from the theorist in an office to the empirical researcher who collects data and formulates theory from it. Furthermore, Grounded Theory places a significant emphasis on the procedural aspect of analysis. The term 'cyclic working' is pivotal here. Grounded Theory is sometimes also referred to as the Constant Comparative Method. This constant comparison is evident in the cyclic collection of data and the cyclical development of analysis, where continuous comparison with the data leads to constant adjustments and refinements of the coding and analysis.

In this section, we focus on the role of theory and the development of theory within Grounded Theory. The more methodological core of Grounded Theory will be discussed in the following section, where we delve deeper into the different steps of the analysis and the role of constant comparison within it.

Charmaz (2006) distinguishes between two perspectives on what constitutes a theory. In positivist approaches to theory, concepts are transformed into variables with operational definitions, and hypotheses are used to test the relationships between these concepts. In contrast, the constructivist or interpretative approach to theory emphasizes its abstract nature and focuses on patterns rather than linear causal structures. Theory is viewed as a blend of values and facts, a historical construct with a processual nature. Strauss and Corbin (1998) define theory as "a set of well-developed concepts related through statements of relationship, which together constitute an integrated framework that can be used to explain or predict phenomena" (p. 15). Charmaz (2006) identifies elements of both views in this definition. The clear focus on concepts aligns with the positivist perspective, while the emphasis on relationships between concepts adds an interpretative component. Depending on the emphasis, one could thus speak of a positivist and an interpretative version of Grounded Theory. However, this book does not delve deeper into this distinction.

At the heart of a theory lie abstract theoretical concepts and the relationships established between these concepts. This raises the question: where do these concepts originate? The standard response in Grounded Theory is: from the data. This approach is known as analytical induction, a concept attributed to Znaniecki (1934). Interestingly, Strauss and Corbin resisted the notion that the foundation of Grounded Theory lies solely in analytical induction. According to Dey (2004), the essence of Grounded Theory was not so much in discovering universally valid truths but in constructing a theory. They also rejected the deductive derivation of hypotheses from existing theories (Glaser & Strauss, 1967). The formulation and testing of hypotheses were contrary to their concept of theory emerging organically from the data. However, the idea of theoretical sampling in Grounded Theory does involve a form of deduction, as it uses existing knowledge to select respondents who can contribute to the research. Consequently, Grounded Theory occupies an intermediate position, emphasizing inductive processes where *sensitizing concepts* make the researcher aware of existing knowledge without hindering the acquisition of new insights from the data.

To extract theory from empirical data, Glaser and Strauss argue that a researcher must possess or develop a certain level of **theoretical sensitivity** (see Glaser, 1978). Theoretical sensitivity refers to the researcher's knowledge and ability to form categories or themes from raw data and to develop dimensions and properties of these. It involves the capability to imbue data with meaning and to conceptualize it into more abstract units. Partly, it is a personal skill of the researcher to think conceptually and derive theoretical insights from qualitative data. This sensitivity can also be honed. Glaser (1992) emphasizes the importance of a sociological or analytical education, where the deductive element of prior theories comes into play again. Through extensive training in social scientific thought, a researcher builds familiarity with theory formation and theoretical thinking. A novice researcher learns to think in terms of causes, effects, contexts, coincidences, correlations, false associations, and so on. Besides this foundational training, literature is also highlighted as a source of theoretical sensitivity. This brings us back to the concept of sensitizing concepts, which guide the researcher's direction in their study and enhance their theoretical sensitivity when examining their data. Theoretical sensitivity is not so much about discovering new ideas or concepts; Glaser and Strauss assume that most ideas and concepts have already been formed in some way. In Grounded Theory, the emphasis is on making new connections between concepts and ideas. Researchers are expected to link the known in new and unexpected ways, which is where the advancement of a Grounded Theory lies.

A Step by Step Example of a Qualitative Analysis with Grounded Theory

Conducting a qualitative analysis typically involves similar steps, yet it is never executed in exactly the same way twice. The essence of this method lies in its cyclical process, involving continuous comparisons and modifications of previously discovered results. This cyclical nature manifests in two distinct phases of the research. The researcher begins by conducting several interviews, for instance. These interviews are transcribed and then analyzed. Based on this initial analysis, the topic list or interview protocol is revised, and new data is collected. This is followed by a phase of in-depth analysis, after which more interviews are conducted. With each round of data collection, the interviewing becomes more focused as is the data collected. The researcher increasingly concentrates on filling the gaps in their analysis or on testing hypotheses that were developed during the analysis.

Furthermore, the cyclical nature of Grounded Theory is also evident in the analysis itself. This analysis operates cyclically, based on the principle of constant comparison. The researcher refines their concepts by continually comparing the results of their analysis with the data, thereby checking the extent to which their theoretical work is valid. This ongoing comparison strongly highlights the analytical induction described earlier: the researcher repeatedly attempts to challenge their findings by incorporating new data into the analysis. The more the researcher's concepts and hypotheses withstand these tests, the more solid the foundation of their theory becomes.

The analysis phase itself can be divided into various sub-steps, each intertwined with one another and with data collection. At the core of the analysis, the researcher first deconstructs the data and then reconstructs it. Deconstruction involves breaking down the vast amount of data typically generated in a qualitative study into smaller units. During this process, some data deemed temporarily redundant is trimmed, leaving only the data relevant to the researcher. In the next step, these data segments (coded references) are interconnected. Themes, categories, or concepts are developed from them. This is the first step in constructing a grounded theory. The second step involves selecting one category as the central category and relating other categories to this central one, thus establishing relationships between various categories. The final phase is writing the entire narrative for the publication.

As mentioned, this process of breaking down (coding the data) and building up (reconnecting the references) is common in most qualitative analysis methods. Within Grounded Theory, specific terms have been assigned to these phases (for each phase, we also provide the original definition formulated by Strauss and Corbin):

- **Open coding**

> The analytic process through which concepts are identified and their properties and dimensions are discovered. (Strauss & Corbin, 1990: 101)

Open coding represents the phase where data is segmented into smaller units. This stage involves assigning names or labels to sections of text within the data. Essentially, you isolate distinct units of meaning you deems relevant for addressing your research question. Some researchers limit the term "coding" to the mere act of labelling raw data. However, in Grounded Theory, this is just the initial phase, and the subsequent tasks also fall under the umbrella of "coding" processes. In NVivo, the term coding is also largely reserved for what Grounded Theory considers "Open Coding" (see Chaps. 8 and 9) .

- **Axial coding**

> The process of relating categories to their subcategories termed "axial" because coding occurs around the axis of a category, linking categories at the level of properties and dimensions. (Strauss & Corbin, 1990: 123)

The outcome of the open coding process is an extensive set of codes (what we will call "a codebook" later). These are numerous labels that might sometimes overlap but have not yet been connected. The task of linking these disparate codes into a cohesive whole is accomplished through axial coding. During this phase, concepts are identified and further developed using the open codes.

- **Selective coding**

The process of integration and refining the theory. (Strauss & Corbin, 1990: 143)

The final step in the process involves interconnecting the concepts. This is where the theory is formed and elaborated: What processes lead to what outcomes? How can one type of variation be understood in terms of another? Typically, in this phase, one concept is chosen as the central category. This represents the most crucial aspect of the theory and is what the researcher uses to address their research question.

To summarize the entire process: The research cycle begins with preparation. Here, the researcher articulates the problem statement, selects a data collection method, and prepares for data collection. This is followed by the first round of data collection. On these data (which, in the case of interviews, are transcribed), the researcher conducts an initial analysis. Through open coding, you start to get a handle on the data. As open coding progresses, axial coding commences. Thus, even before collecting new data, the researcher gains partial insight into the emerging concepts. Then, a second round of more targeted data collection follows. The same analytical process is repeated, with the researcher coding the new data (i.e. open coding) and further developing the (axial) codes formed before. The principle of constant comparison is crucial here, as the development of codes involves continual reference back to the original data. The evolving theory is derived from but grounded in the data. This cycle may repeat several times, leading the researcher into the final phases of the study. You subject your developed axial codes to selective coding, elaborating the relationships. Additional data may be collected during this phase if needed. Finally, the researcher writes up the analyses and possibly presents them to the respondents from whom the data were collected (peer debriefing).

References

Bong, S. A. (2002). Debunking myths in qualitative data analysis. *Forum: Qualitative Social Research*, *3*(2), 1–12. http://www.qualitative-research.net/fqs/fqs-eng.htm

Charmaz, K. (2006). *Constructing grounded theory*. Sage.

Cresswell, J. W. (1998). *Qualitative inquiry and research design*. Sage.
Dey, I. (2004). Grounded theory. In C. Seale, G. Gobo, J. F. Gubrium, & D. Silverman (Eds.), *Qualitative research practice* (pp. 80–93). Sage.
Glaser, B. G. (1978). *Theoretical sensitivity: Advances in the methodology of grounded theory*. Sociology Press.
Glaser, B. G. (1992). *Basics of grounded theory analysis: emergence vs. forcing*. Sociology Press.
Glaser, B. G. (2001). *The grounded theory perspective I: Conceptualization contrasted with description*. Sociology Press.
Glaser, B. G. (2002). Constructivist grounded theory? *Forum: Qualitative Social Research, 3*(3), 1–10.
Glaser, B. G. (2003). *The grounded theory perspective II: Description's remodeling of grounded theory*. Sociology Press.
Glaser, B. G. (2005). *The grounded theory perspective III: Theoretical coding*. Sociology Press.
Glaser, B. G., & Strauss, A. L. (1967). *The discovery of grounded theory: strategies for qualitative research*. Aldine Publishing Co.
Moustakas, C. (1994). *Phenomenological research methods*. Sage.
Richards, T. (2002). An intellectual history of NUD*IST and NVivo. *International Journal of Social Research Methodology*, 5(3), 199–214. https://doi.org/10.1080/13645570210146267
Spradley, J. P. (1980). *Participant observation*. Holt.
Strauss, A. L., & Corbin, J. (1990). *Basics of qualitative research: Grounded Theory procedures and techniques*. Sage.
Strauss, A. L., & Corbin, J. (1998). *Basics of qualitative research*. Sage.
Tesch, R. (1990). Qualitative research: Analysis types and software tools. *The Falmer Press*. https://doi.org/10.4324/9781315067339
Znaniecki, F. (1934). *The method of sociology*. Farrar & Rinehart.

CHAPTER 3

A Quick Tour of NVivo

Key messages in this chapter

- We explain what a program like NVivo does.
- We give a short intro on the essence of code-and-retrievieving data.

What Can NVivo (Not) Do for You?

The development of software to aid with qualitative research started in the early eighties of past century and saw a huge diversity of programs all promoting some specific advantage over the other (Tesch, 1991; Weitzman & Miles, 1995). But as happens in many new markets, mergers and takeovers lead to concentration of the market until only a few players remain. Today, NVivo is one of the dominant players in the market of CAQDAS, together with Atlas-ti and MaxQda. NVivo is a software program to perform Computer Assisted Qualitative Data Analysis (CAQDAS). The software is the successor of the NUD*IST program developed in 1981 by Richards (2002) in close collaboration with Richards (1987).

Later in this book, we will delve deeper into NVivo's three fundamental attributes: data management, coding data, and analysing data through querying coded information or formulating conceptual frameworks. These attributes are the program's core tools in a qualitative data analysis process. Post data collection, NVivo facilitates the transcription of interviews or group discussions (even though using automated AI-transcription could be more

D. Mortelmans, *Doing Qualitative Data Analysis with NVivo*, Springer Texts in Social Sciences, https://doi.org/10.1007/978-3-031-66014-6_3

efficient for you today). Next, NVivo is also an all-encompassing data management system. All material gathered throughout the project can be stored in a single project file (the nvp-file). The program is very flexible in the way the material is organized. Users can create folders and make hierarchies in their storage as they like.

Following data import and organisation, the data can be coded. Coding serves as a data reduction tool in order to structure the data in smaller entities (called references). Coding is not only limited to a text-labelling process. Also identities and units (interviewees, pictures, focus groups, movie characters) can be identified to make comparisons later on more easy. In Classifications, users can bring together properties of these units and code the raw material with them. When the raw material is coded, a codebook is available and the data is broken down in smaller pieces organized according to the codes attached to them.

At this point, the researcher needs tools to bring together the coded material and to look at the regularities and irregularities in the data (outliers are in qualitative research just as important as general trends). The instrument to perform these searches in the coded data is the Query Tool. In fact, we should talk plural since NVivo has no less than eight different types of queries on board. These different queries allow to perform very specific searches. The output of a query is a collection of references on one theme (the code or codes being selected).

From that point onwards, NVivo reaches the limit of what it can do for the researcher. NVivo does not produce automated theoretical schemes or storylines (see Fig. 3.1). You get a list of references that were scattered around your data before and presented now together in one results window of a query. You start reading and interpreting these results, compare them across different groups and backgrounds, and write about the meaning of what respondents have spoken about or what visuals communicate.

Fig. 3.1 Tools inside and operations outside NVivo

INSIDE NVIVO	Data management Coding data Querying (retrieving) coded data
OUTSIDE NVIVO	Reading Interpreting Writing

Code and Retrieve as the Fundamental Principle

The previous paragraph makes clear that, historically, qualitative software programs have always focused on a 'code-and-retrieve' approach. Researchers import raw data into the software and somehow will 'code' their data. Subsequently, the software would facilitate the retrieval of this coded material via queries or a code retrieval tool. Even though decades have passed, the fundamental principle of coding remains largely unaltered. Coding continues to be a pivotal element of NVivo, and the software is designed to enable users to link their codes to their data as effortlessly as possible. We will show this in detail in Chaps. 8 and 9. The nature of the raw material—whether text, video, audio, or graphics—is unimportant. NVivo is capable of coding a variety of data types (see for example Chaps. 16 and 17).

As coding is such a crucial core element, NVivo will make it users as easy as possible to attach their material to Codes (or vice versa). There is the visual aid, called Coding Stripes at the side of the screen, show which codes have been attached to a fragment. With coloured bars, the researchers retains a clear overview of codes used and overlapping codes at fragments. Next, the Highlighting function gives a coloured background to coded material (most visible in text material). With this background colour, it immediately is clear which material is coded and where fragments are identified. The coding stripes are also used later when overviewing results from Queries (more information in Sections "Coding Stripes" and "Code Highlighting" in Chap. 8).

All codes are collected in a codebook. This is a section in the project where every code in the project is stored. As the coding process evolves, users can create hierarchies in their codebook by organizing the codes in trees. Some codes serve as Parent Codes and others as Child Codes (see Section "Making Code Hierarchies in Your Codebook" in Chap. 8). Crucial in this organisation is the Aggregate-option. Aggregating coding allows users to push coding from lower (child) codes up in the tree to the higher order (and often more abstract) codes. This aggregation makes grouping codes easier since coding only needs to be done at a low level, close to the raw material. Any codework that eventually end up in a hierarchy can easily be distributed along the tree (see Section "Aggregating Coding Work" in Chap. 8).

Coding is a manual work and is considered as one of the most laborious phases during the qualitative analysis. The coding process is made as easy and simple as possible with one central Coding screen where new codes are created and existing codes are attached to references. Also drag-and-drop coding speeds up the coding process when only a limited number of codes are involved. Queries can also be used as a coding tool. E.g. you can use the Text Search Query (see Section "The Text Search Query" in Chap. 12) to locate instances of particular texts in the material and consequently code these references. Artificial Intelligence tools like machine learning allow users to autocode their data. As a first round of coding researchers can have themes or sentiments identified in their data and build upon these codes in the later coding process. Also machine learning allows researchers to have new data coded, based on existing coding work present in your project (see Chap. 19).

Queries (Retrieval of coded material)

For a long time, searching for text was one of the essential parts of the initial qualitative software package. The code-and-retrieve logic was a breakthrough in the second generation software and excellence of this type of software was often measured in terms of how good is the coding ***and*** how powerful is the retrieval? The query tool is NVivo's retrieval tool after the coding of the raw material. In fact, there is not one query tool but eight. For specific search operations, NVivo offers the users different queries. Most of the queries require your data to be coded but not all of them.

The *Coding Query* is the basic code retrieval query (see Section "The Coding Query (Some More Details)" in Chap. 12). As a start, users can search for coded material in the project. NVivo returns a set of fragments (with coding stripes) so that the researcher gets an overview of all instances where this theme has been discussed in the project. To avoid context-detached analyses, NVivo offers the possibility to expand the results of the search to a broader context than just what has been coded. As such, researchers that code on a word-basis are always sure they can see the surrounding sentence or paragraph in which this word appeared. Never ever should the analysis be done outside the context of where themes have been coded (see Section "Dissecting the Query Window" in Chap. 12). The coding query starts with a simple search for one code but can easily be expanded to more advanced searches using Boolean logic and other search options.

The *Text Search* Query allows to search for instances of text in raw text material. Again, Boolean logic and search options allow the researcher to specify in great detail what kind of text needs to be searched for. The Text Search query is also suited to be used as a coding device (see the previous paragraph). As long as specific terms in the texts can be identified, the text search query can find them and code them. As such, some recurring terms in your material can be coded even though it will never fully replace the manual coding of the raw material.

The *Matrix Coding Query* (see Section "The Matrix Coding Query" in Chap. 12) is the most powerful tool in the program. It allows to combine a search of some Project Items in rows with a search of other Project Items in columns. The Matrix Coding query allows complex combinations not only of two (sets of) codes but allows also to cross attributes from a classification or even the work of users (to allow codebook comparisons). The possibilities of the Matrix Coding query are endless and each query results in a matrix where the cells show the combination of material in the rows with material in the columns. When double clicking on a cell, the researcher is shown all fragments that are available at that intersection. As such, it is very easy to make comparisons in the project and gain a full understanding of what is going on in the project. An extension of the Matrix Coding Query is the Crosstab query. This is a variant of the Matrix Coding Query but with the possibility to crosstab two attributes from a classification (see Section "The Crosstab Query" in Chap. 12). As such, you can make comparisons at the intersection of age and gender (young women vs. older women) where the Matrix Coding Query only allows to compare values of gender or values of age.

References

Richards, L, & Richards, T. (1987). Qualitative data analysis: Can computers do it? *Journal of Sociology*, 23(1), 23–35. https://doi.org/10.1177/144078338702300102

Richards, T. (2002). An intellectual history of NUD*IST and NVivo. *International Journal of Social Research Methodology*, *5*(3), 199–214. https://doi.org/10.1080/13645570210146267

Tesch, R. (1991). Computer programs that assist in the analysis of qualitative data: An overview. *Qualitative Health Research*, *1*(3), 309–325.

Weitzman, E. A., & Miles, M. B. (1995). *Computer programs for qualitative data analysis: A software sourcebook.*

CHAPTER 4

Getting Started 1: Installing and Configuring NVivo

Key messages in this chapter

- NVivo is installed through the standard installation wizard in Windows or the app on MAC.
- The Microsoft SQL Server is crucial for NVivo to work.
- The User Profile is essential to identify you as a user in your project(s).

Installing the Program

NVivo is distributed through individual licenses you buy on the website of Lumivero or via an institutional license. When you buy the software privately, you need to create a login on the NVivo website, order and pay for the software and then download the software from your account. When your workplace has an institutional account, they will distribute the set-up file internally.

When you start the set-up, a set-up wizard appears, asking you step by step to provide details on the installation process. We only repeat some steps in this paragraph, as most are straightforward.

In Fig. 4.1, we highlight two steps in the wizard that might raise specific questions. First, NVivo wants to install an additional program called *Microsoft SQL Server*. This is a component that NVivo uses in the background, and it is *crucial* for its functioning. Therefore, you need to install this program

D. Mortelmans, *Doing Qualitative Data Analysis with NVivo*, Springer Texts in Social Sciences, https://doi.org/10.1007/978-3-031-66014-6_4

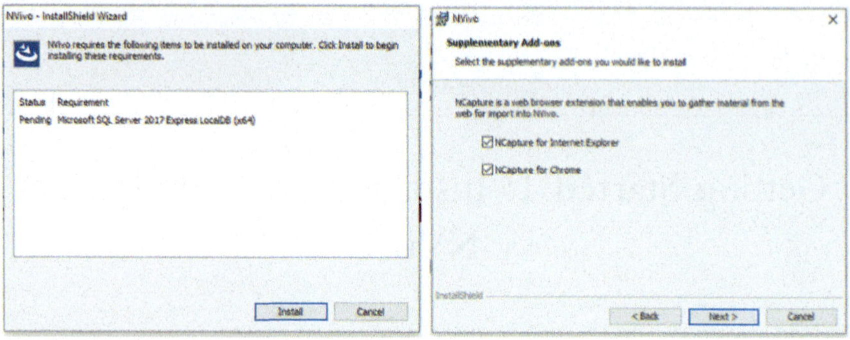

Fig. 4.1 Required (left panel) and optional components (right panel) of the NVivo installation

together with NVivo to get the program working. Second, NVivo asks whether you want to install the free add-on NCapture. This is only crucial if you want to work with Social Media material in NVivo. As we will explain in Chap. 17, NCapture is an essential tool for importing websites, Facebook threads, or tweets into NVivo. If you do not plan to work with this kind of material, you do not need to install this add-on.

> **For the Mac user**
>
> - For MAC, the installation file is a.dmg file that can be used in the standard installation application on your computer.

Registering the Program

Once the program is installed, you must register NVivo to use the software. The registration process depends on the type of license you have. If you have an individual license (you bought the program online), you need to log into your NVivo account for the program to check your license and give you access to the program. If your institution bought a volume license, they must provide you with their institutional key to use the program (see Fig. 4.2). You can use the program for free in a 14-day trial if you have neither. Once you have identified your license type, NVivo will present a *License Activation* window asking for details about your user profile. When you submit the form, the message "*License has been activated*" appears, and you can start using the program.

After the license is activated and you want to use the program, NVivo will ask you to create a user profile. NVivo is a database program that logs all user activities in the background. To know which user performed which action, NVivo needs to know who uses the program. Therefore, the software asks

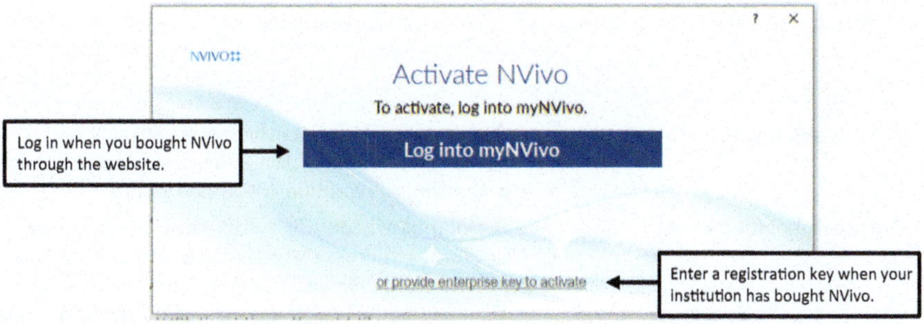

Fig. 4.2 Registration of NVivo

you to fill in your name and initials. When you start working with your tools and create new project items, you will see that NVivo shows your initials as a sign that you, as a user, have created or changed a particular project item.

Updating the Program

Right after your installation or later on regular occasions, NVivo will ask you to update the program. These updates are essential as they often add new tools to the program that were unavailable when a major version update was launched. So, if you want to use the full potential of your version or increase stability, it is important always to install the updates.

Important to know is that sometimes the file format of your projects changes not only between versions but also within versions. In that case, the update will cause NVivo to convert your project file to the new file version, making it unreadable for previous installations of that version. Therefore, it is essential for collaborative projects to always update to the same version to exchange projects within the project team (see also Chap. 19).

Remark

- Your license determines the updates you are entitled to. When you buy a license on the website of QSR, you buy a license for a particular version of NVivo (e.g. version R1), and you are entitled to updates of that specific version only (e.g. updates from version R1.0 to R1.6). When your institution buys a volume license, this usually is a yearly license entitling all users to get *any* update throughout that year. That means that when NVivo upgrades from version R1 to version 14, you are entitled to install the 14 version when it is released. For the other license type, you will be asked to pay an upgrade price to move from version R1 to version 14.

Table 4.1 Overview of add-ons in NVivo at additional charges

Tool	*Explanation*
NVivo transcription	A transcription service where you can buy transcription of your data at an hourly basis or with a monthly subscription (50 h per month)
NVivo collaboration cloud	A cloud-based tool for collaborative projects that does not require a server installation but that puts projects in the cloud. It is not available for mixed Windows and MAC teams. The team needs to work in one of the two operating systems

Installing Add-Ons

When you install NVivo, you get a fully functional version of NVivo with all the tools you need for analysing a qualitative research project. But on top of the basic functionality, Lumivero (the firm that develops and sells NVivo) also created some additional add-ons that you can order for an additional cost. In this paragraph, we only give a listing of these add-ons. We will not discuss any of them in this book (Table 4.1).

Updating Your License

Some users use NVivo in a volume license. That means that an institution (e.g. a university) pays a yearly fee for its staff and students to let them use the program in their work. Consequently, these users need to renew their license yearly as institutions cannot buy a perpetual volume license but only a yearly one. When the deadline arrives to renew the license, users get the message to renew their license (see Fig. 4.3).

Go to File > Product Info > Manage License > Extend License to renew the license. You get the *Extend License* window to enter the new license key.

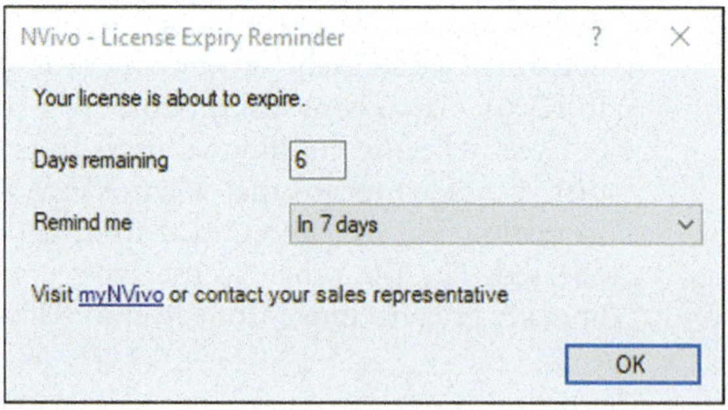

Fig. 4.3 Update of the yearly license of NVivo

CHAPTER 5

Getting Started 2: The Workspace and Jargon of NVivo

Key messages in this chapter

- The workspace of NVivo resembles that of Microsoft Outlook.
- Some tabs in the ribbon are mirrored in the blue *Navigation View*.
- Most jargon in the program is self-explanatory.

Main Layout of the Workspace

In the next chapter, we will discuss the start of your workflow (starting a new project and entering the program). In this chapter, we give you an overall introduction to the main user interface of the program and the different areas on your screen that determine the overall workflow when you work with NVivo. In this book, we will only present the English version of the user interface. In **File > Options > General**, you can change the user interface language. The NVivo interface is available in English (default), French, German, Spanish, Japanese, Portuguese, or Simplified Chinese.

For the Mac user

- The MAC version is not yet available in Portuguese or Simplified Chinese.

D. Mortelmans, *Doing Qualitative Data Analysis with NVivo*, Springer Texts in Social Sciences, https://doi.org/10.1007/978-3-031-66014-6_5

In Fig. 5.1, we show the three main areas in the interface. The blue zone at the lefthand side of the screen is called the *Navigation View*. Here, you find all the main components of an NVivo project: Data, Coding, Cases, Notes, Sets, Queries, Visualisations and Reports. Each of these sections is subdivided into predefined folders. For example, in the section Data, you'll find three folders: Files, Files Classifications and Externals. These folders are always present in a project. They cannot be removed or renamed. However, a user can create subfolders in several of the pre-defined folders (not in all). For example, in the example project, the pre-defined folder Files is further subdivided into six custom-made folders.

You can find the *List View* and *Detail View* window on the right side of the screen. The *List View* shows the content of a folder (that you have selected in the *Navigation View*). All project items in a specific folder are listed with extra information columns like the creation date and the user that created or changed the project items. The *Detail View* window is only visible when a project item is opened or when a new project item is made. Opening project items can be done by double-clicking them (e.g. opening files). Some new project items, like queries, are created and defined in the *Detail View* window. Both windows operate differently. The *List View* only shows the content of *one* selected folder. Selecting a new folder in the *Navigation View* changes the content of the *List View*. On the other hand, the *Detail View* window is a set of different tabs stacked horizontally (see Fig. 5.2). Multiple tabs can be opened simultaneously, and the user can switch between the tabs by clicking on the name of the tabs. Tabs can also be closed by clicking the ⊠ next to the tab's name. When you close the last tab, the *Detail View* window disappears, showing only the List View.

In Fig. 5.2, some more details on the user interface are highlighted. First, NVivo has a ribbon like most Microsoft Office programs where the options of a particular menu are shown with icons. But in some tabs in the *Detail View* window, NVivo presents a second, smaller ribbon with options specific to that

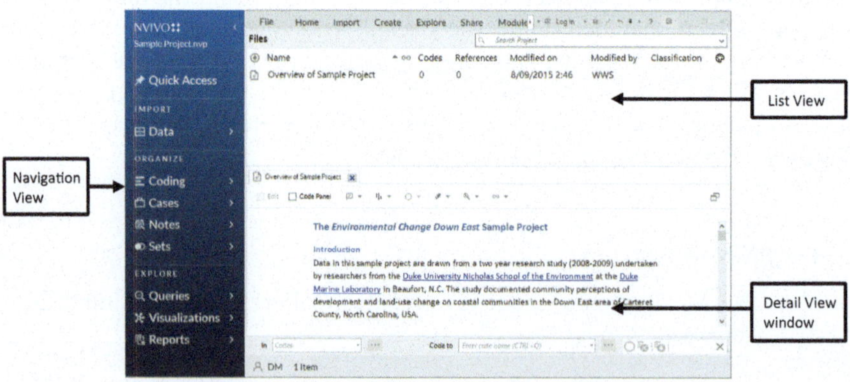

Fig. 5.1 Main interface of NVivo—part 1

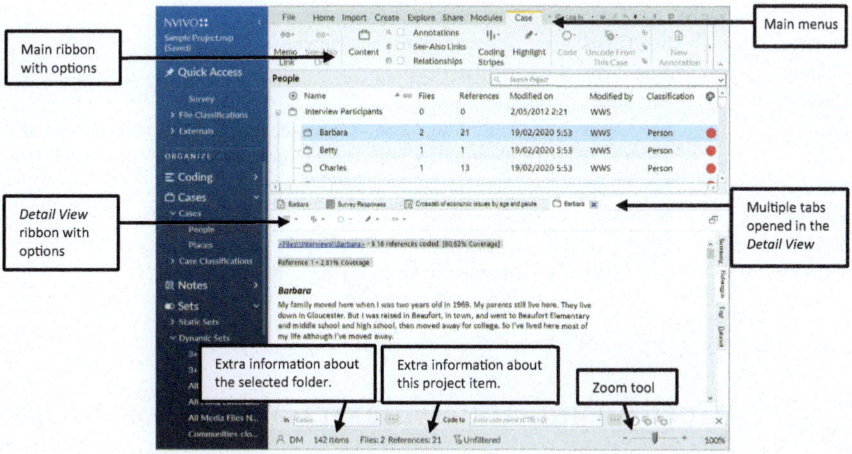

Fig. 5.2 Main interface of NVivo—part 2

project item. For example, when you open a file, you get coding icons in the ribbon of the *Detail View* window that help you with coding the file. Not all tabs in the *Detail View* window will have a second ribbon. For example, query definitions will not have their own *Detail View* ribbon.

At the bottom of the screen, NVivo shows some summary information about the *List View* and the *Detail View* window. NVivo reports the number of items saved in the selected folder from the List View. From the *Detail View* window, you get some details about the project item opened in the *Detail View*. The example in Fig. 6.2 shows that the selected folder contains 142 project items. From the (case of the) interview of Barbara selected in the *Detail View* window, you learn that this interview is coded with two files and contains 21 text references. In the bottom right of the screen, you find a zoom tool that enables you to zoom in or out on the information shown in the *Detail View* window.

The Navigation View

In this paragraph, we take a closer look at the different components of the *Navigation View*. The *Navigation View* is the central hub of a project in NVivo, and the researcher will switch between its components when working on different elemen*ts during their analysis.*

In Fig. 5.3, we present a commented overview of all sections in the Navigation View. You can see these different sections as separate tools to work with your data. Depending on the stage of your analysis, some sections will be used more intensely than others. As it is too early in this book to go into detail on all these tools, we refer you to the chapters where we will extensively discuss the tools you find in the *Navigation View*.

Data	See Chap. 6 on primary data management
Coding	See Chap. 8 on thematic coding
Cases	See Chap. 9 on case coding
Notes	See Chap. 7 on memoing, Chap. 11 on secondary data management and Section "Making Summaries with the Framework Matrix" in Chap. 10
Sets	See Chap. 11 on secondary data management
Queries	See Chap. 12 on querying
Visualisations	See Chap. 13 on visualisations with maps
Reports	See Chap. 14 on reporting and exporting data

Customising Your Workspace Outlook

To conclude this chapter, we want to show two customisations that help you to gain efficiency when working with project items. When you double-click on a project item (e.g. the interview with Barbara in Fig. 5.4), the *Detail*

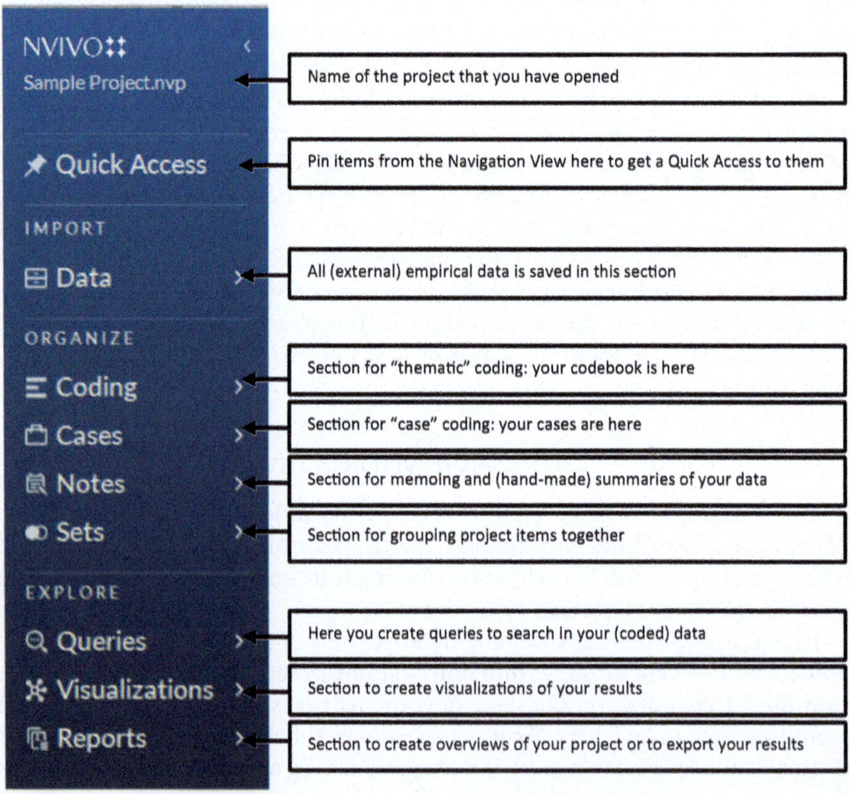

Fig. 5.3 The *Navigation View*

View window will open, and the document will be shown in a separate tab. By default, the *Detail View* window is opened underneath the *List View*, as is shown in the left panel of Fig. 5.4. This is called the *bottom view* of the *List View*. In **Home > Workspace > Right,** this view can be changed in the *right-hand view* where the document is now placed at the right-hand side of the *List View*. This has the advantage of seeing more items in the List view on the left side and reading more of the interview in the *Detail View* window on the right side.

For researchers with two screens, you can also undock the *Detail View* window from the main application. As such, you can put the *Detail View* window on your second screen to utilise all the available window space. You undock the screen by **Home > Workspace > Undock** (as shown in Fig. 5.5). When you want to integrate both windows again, you go to **Home > Workspace > Dock** in the window that contains the *List View* content.

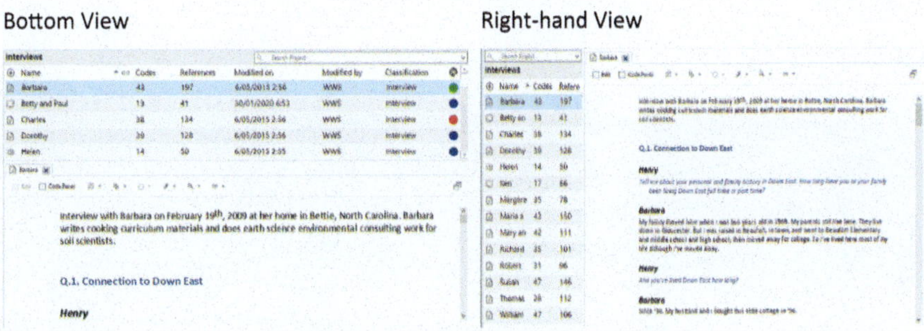

Fig. 5.4 Position of the *List View* and *Detail View* window

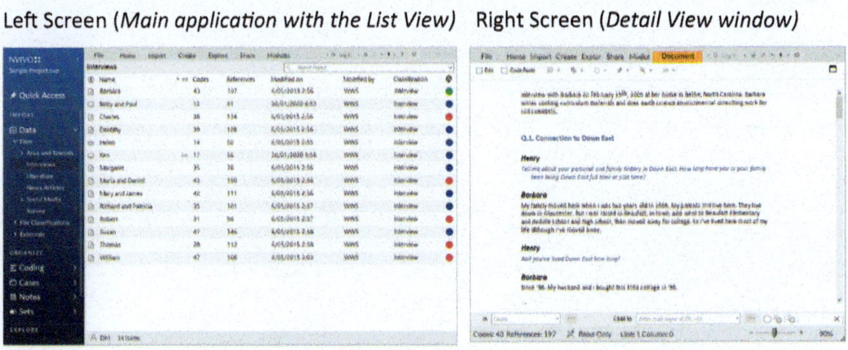

Fig. 5.5 Undocking the *Detail View* window

CHAPTER 6

Setting Up Your Project: Primary Data Management

Key messages in this chapter

- Your project is saved in ONE file:.nvp.
- Make regular backups so you don't lose work should the program crash.
- NVivo has a pre-prepared project structure that users customise.
- All data is imported into the project file or created inside the project file.
- NVivo allows you to transcribe audio and video.

Starting or Opening a Project

An NVivo project includes all research data that you enter into NVivo and all new data and analyses that are created on top of it. In other words, a project contains all project items needed to perform a qualitative analysis, from data over codebooks to visualisations. Each NVivo project is saved as one single file with a.nvp extension.

For the Mac user

- Your project is saved as a.nvpx file.

D. Mortelmans, *Doing Qualitative Data Analysis with NVivo*, Springer Texts in Social Sciences, https://doi.org/10.1007/978-3-031-66014-6_6

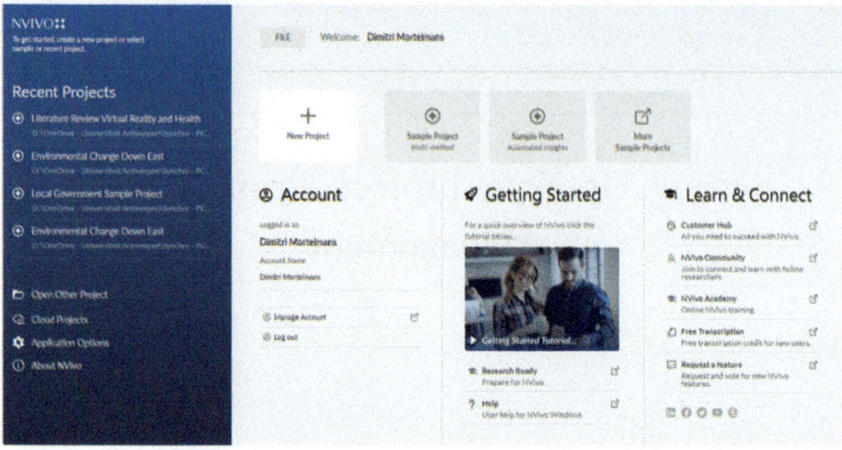

Fig. 6.1 Welcome screen of NVivo

When you start the program, you do not immediately get access to the entire interface of the program. Instead, the *Welcome* screen appears (Fig. 6.1):

To start a new project, click on the *New Project* button. If you have already worked with an NVivo project, you will find a list of recently used projects in *Navigation View* > **Recent Projects**.

The *Multi-method* and *Automated Insights* sample projects are available from the *Welcome* screen. The other sample projects are available online through *More Sample Projects*. This button opens a browser and goes to the Help Pages of NVivo, where you can download more sample projects from the NVivo website. The following projects are available online:

- Literature review: Virtual Reality and Health (see Chap. 21).
- Mixed methods: Wellbeing in the Older Women's Network (see Chap. 18).
- Survey: Top High School Alumnae (see Chap. 18).

When you create a new NVivo project, NVivo asks for information in a two-step wizard. In the first step, you need to provide a title for your project and determine where to save the project file. By default, this will be your standard *Documents* folder. With the *Browse* button, you can change the default location to save your project. You can also activate the user's log with ☑ *Keep a log of user actions*. This log will allow you to retrace your analytical steps during the analysis. When activated, you can always consult your log through **File > Project Information > Open Project Event Log** (Fig. 6.2).

In step 2, you set up the Autosave feature and the Project Recovery File. Important is the choice between autosaving the project (suggested time is 15 min) and the Undo function. Within NVivo, unfortunately, you have to

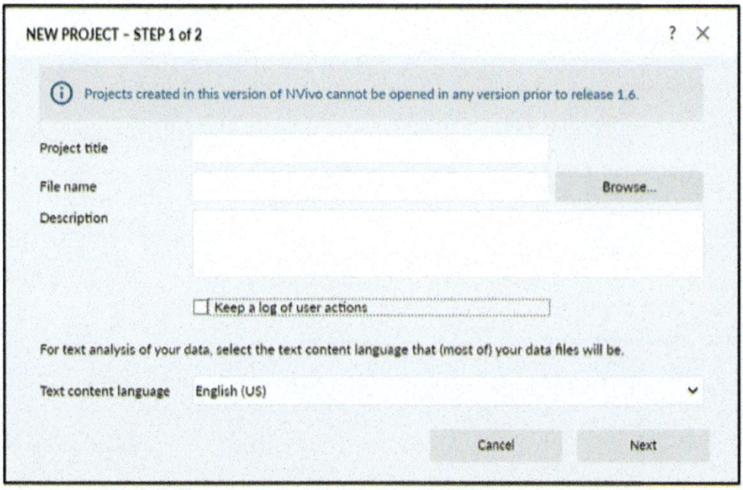

Fig. 6.2 Creating a new project—STEP 1: general project information

choose between autosaving and the ability to use the 'undo' function. We recommend to forego the automatic autosaving as the 'undo' function is something most users will want to use regularly. Since the Undo function helps users more during their workflow, we recommend not switching on the Autosave function. Of course, as a user, you will need to save your project regularly on your initiative to avoid losses in case of a crash. The lower part of the window lets you choose the options for the Project Recovery file. This recovery file is created in a folder *NVivo Recovery* in your main *Documents* folder. Beware that this is only a recovery file and not a backup (see next paragraph on making backups). The recovery file helps restore crashed projects but cannot be opened as a stand-alone project file (Fig. 6.3).

When you click **Create Project**, the project file is created, and NVivo offers a Tour of the program. You can follow the tour or skip it and start working on your project. The NVivo workspace is now finally revealed, and you can start working. The layout of the workspace was already discussed in Section "Main Layout of the Workspace" in Chap. 5.

Backing Up and Repairing Your Project

When we discussed the two-step wizard in the previous paragraph, in tiny letters, NVivo gave us a warning to make regular backups manually.

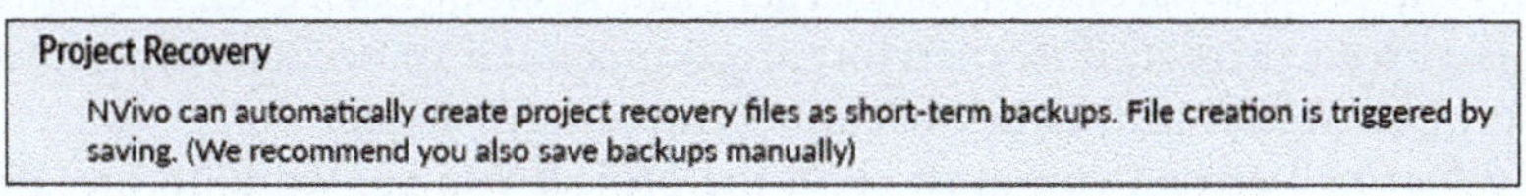

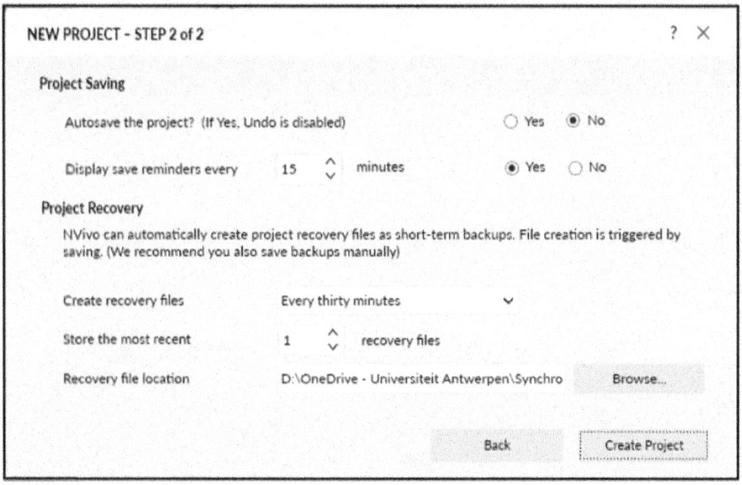

Fig. 6.3 Creating a new project—STEP 2: saving and recovering projects

As we explained before, the recovery file is not a real backup, so you need to take care of your backups of the project file. Backups are important as NVivo might crash occasionally during your analyses. The tiny letters in the Wizard remind us that no automatic backups are made and that the user is responsible for making backups of their project files. You can make backups with **File > Copy Project**. Give a new name to the backup and save it, preferably on a separate medium from where you work (e.g. an external hard drive or in the cloud). When you work intensively with the project (e.g. when you are coding your material), we recommend making a backup daily to separate files (i.e. five backups a week) and keeping one backup file every week (i.e. keep one of these five daily backups and delete the other ones). As such, you can always go back to the previous day and, if necessary, to the previous week(s). This *Copy Project* option also lets you convert your project file from Windows to other formats (Mac and Server).

Sometimes, the project you are working on becomes slow, or elements in the project no longer work properly. This especially happens when you work on the same project for a long time. The program then saves so much material in the same file that, occasionally, things get saved incorrectly. NVivo can screen projects for errors and fix many errors itself if necessary. Along with the error screening, NVivo will compress the project to take up less space. After all, the software uses quite some disk space for its project files so that after a while, the project becomes much larger than it should be. Even without any errors in your project, it is, therefore, recommended to occasionally compress the project so that it does not become unnecessarily large.

To fix small errors, you must close the project you are working on (**File > Close**). You now return to the *Welcome* screen. Even if no project is active,

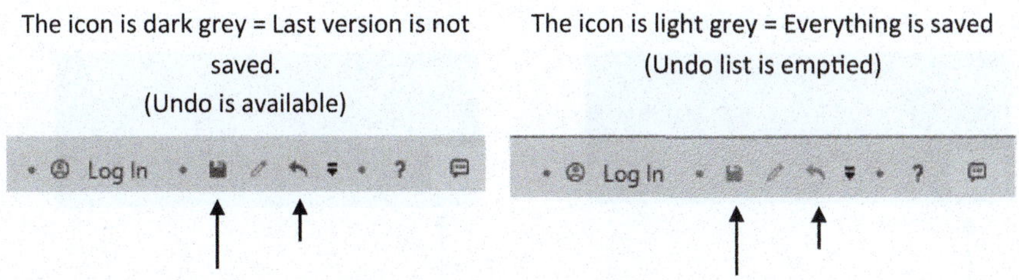

Fig. 6.4 Visual aids in the save button

you will still see some menus at the top that remain available. In **File > Help > Compact and Repair Project,** you can select your project file from where you saved the project and let NVivo repair and compress it. You will receive the message 'Compact and repair operation completed' when the screening is completed. You can now open your project again and continue working.

While you work, NVivo will also periodically ask you to save your project. By default, NVivo recommends saving the project every 15 min (if you haven't already done so). You can adjust the interval at which NVivo asks for this under **File > Project Properties > Save and Recovery**. Adjust the time in the option "*Display save reminders every 15 min*". Again, this option only warns you to save the current project regularly. Also, saving the project does not make automatic back-ups.

For the Mac user

- No *Save reminder* option is currently available.

Whether you get this reminder or not, saving your project is done in **File > Save** or in the top right corner of the program, where you find a small *save* icon. Both ways to save changes are only available if there are changes in the project to be saved. The colour of the icon immediately shows whether there is unsaved material in your project or not (Fig. 6.4).

Setting Up a Folder Structure

When you create a new project, NVivo provides a pre-structured Project File. In the *Navigation View*, you will see the different main sections (e.g. Data, Coding, Cases) underneath their predefined folders (e.g. in Data: Files, File Classifications, Externals). These main folders *and* their subfolders cannot be removed or renamed. NVivo uses arrow icons to (1) indicate whether subsections exist and (2) to open and close these subsections (Fig. 6.5).

You can always create your folder structure underneath the standard folders to customise how you want to organise your Project File. To make new

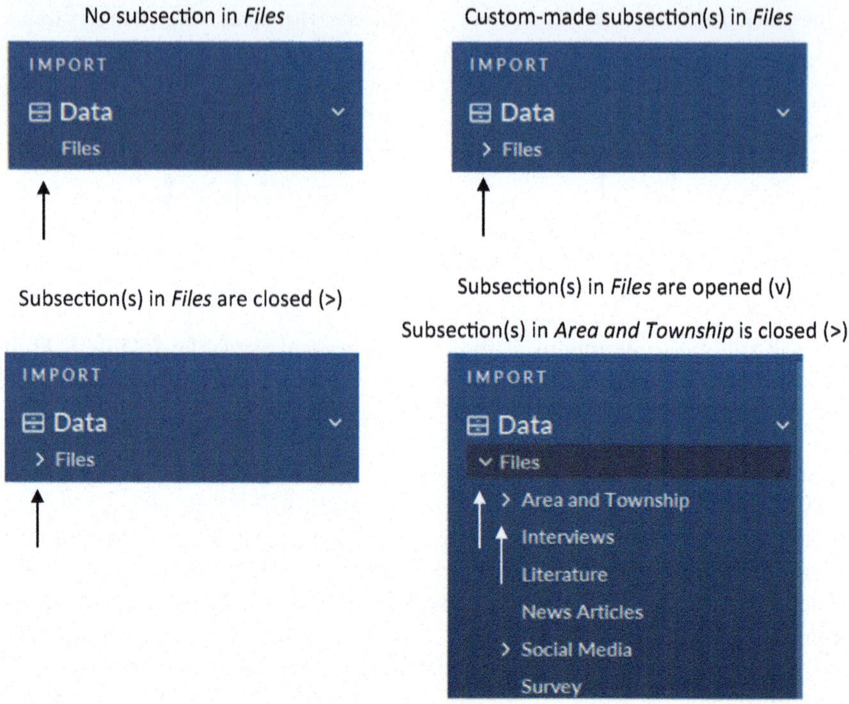

Fig. 6.5 Visual aids on the status of (sub-)folders

folders, select the higher-level folder that will be the main folder. Then, make the folder in **Create** > **Folder** or **RMC** (*above the main folder*) > **New Folder**. Both the built-in folders and the custom-made folders can have further subfolders.

Importing Primary Data in Your Project

An NVivo project file can combine many different sources or files into one NVivo project file. That means that you need to import all your primary material that is made outside the program into the program. The **Import** menu contains all items that can be imported into your project file.

The main types of material most researchers will have available are textual documents (e.g. interview transcripts), images, audio files or video files. All these types of files are imported with **Import** > **Files**. You can select one file or a group of files (not necessarily of one type). The program will offer an option ☑ *Create a case for each imported file*. We discuss this option in Chap. 9 (see Section "Classifications When Importing Data" in Chap. 9). The imported material is placed in **Data** > **Files** in the *Navigation View*. If you want your material in a custom-made subfolder of **Files**, then choose this subfolder first before importing your material. Audio and video files often have a considerable size. To protect the integrity of the project file, NVivo will only import audio and video as embedded files up to 20 MB. You can extend this size to 40 Mb in **File > Options > Audio and Video**.

For the Mac user

- Importing a mixture of file types all at once is not possible. You must import textual documents, PDFs, audio files or video files separately.

Aside from textual data and audio and video files, the ribbon in the **Import** menu shows many other items that can be imported. Most of these items will be discussed in other chapters:

Project	See Chap. 19 on working in teams
NCapture	See Chap. 17 on working with social media
Survey	See Chap. 18 on doing mixed methods
Classifications	See Chap. 9 on working with classifications
Bibliography	See Chap. 21 on doing a literature review

Finally, NVivo will also import special textual data like Notes (from Evernote or OneNote), emails (from Outlook), Codebooks and Reports (from other NVivo projects). Unfortunately, NVivo cannot import Codebooks from Excel or Word. Many researchers have their codebooks in either a word processor or a spreadsheet, but no import is available for these file formats. The Codebook import requires a Codebook from another NVivo project or another QDA program (e.g. Atlas-ti or MAXQda) in the REFI-QDA format. More information on this format can be found in Section "Export Parts of a Project".

Operating Files in Your Project

The imported files are saved to a folder and shown in the *List View*. You can read documents at any time by double-clicking on them. The text will then be visible in the *Detail View* window. If you open multiple documents, you can jump between the different documents via the tabs at the top of the *Detail View* window. When you open a textual document, an extra *Document* menu appears with specific Ribbon options to work with this textual file. Many of these options are discussed throughout the book. Beware, however, that this extra menu only appears when the tab of a textual file is selected in the *Detail View* window (Fig. 6.6).

Similarly, you can open PDF files (with an extra menu PDF), audio files (with extra menu Audio) and video files (with a Video menu). For audio and

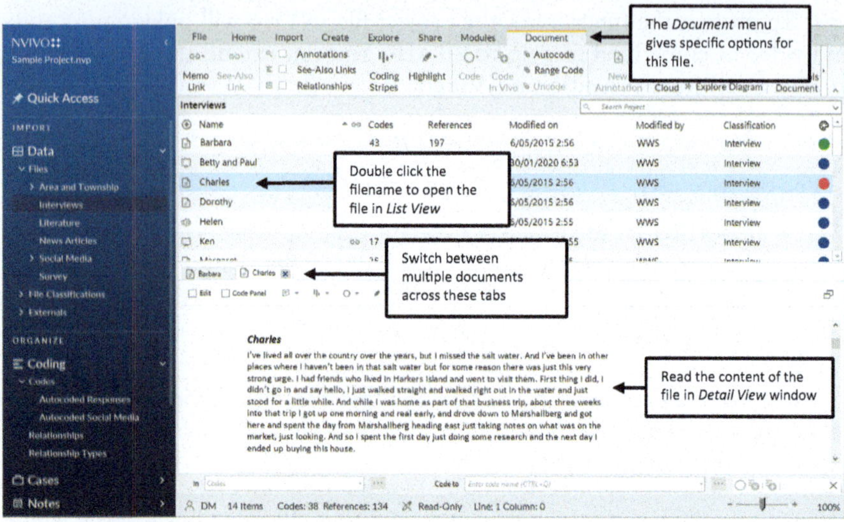

Fig. 6.6 Handling textual files

video files, the extra menu gives you the ability to play the file, to rewind or fast forward and to pause. Also, the volume and the speed at which the file is played can be adjusted. For audio files, you get a visual on the audio trail. For videos, both the audio trail and the actual video are shown (see Fig. 6.7).

When you want to close a file, click on the ⊠ at the right side of the file tab in the *Detail View* window. All tabs have this ⊠ and can be closed separately. When the last Project Item in the *Detail View* window is closed, you see only a *List View* and no longer a *Detail View* window.

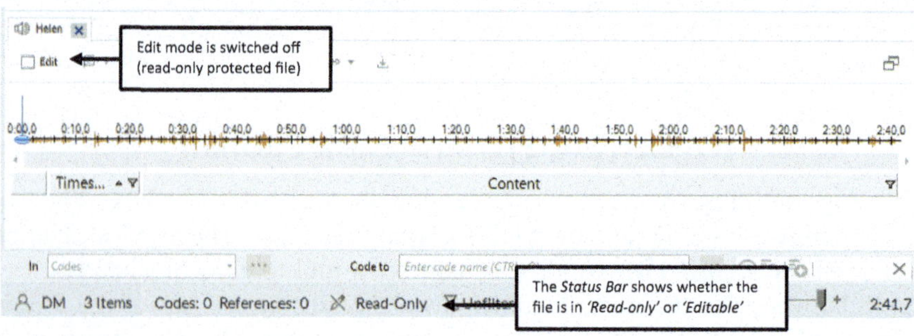

Fig. 6.7 File in protected read-only mode

Transcribing Media Files

In the previous paragraph, we explained how to import material produced *outside NVivo.* In this paragraph, we explain how documents are made in the program and how you transcribe audio (and video) material with NVivo.

This is the overview of the transcription process:

Prerequisites before you start

- The audio or video file you want to transcribe needs to be imported into your project file and opened in the *Detail View* window.

Workplan

STEP 1. Put your media file in *Edit* mode.
STEP 2. Put NVivo in *Transcription* mode.
STEP 3. Make a first *Transcript Entry.*
STEP 4. *Repeat* and make subsequent Transcript Entries.

Step 1. The Edit Mode

When you import files in your project (see Section "Importing Primary Data in Your Project"), NVivo protects these files from accidental editing. When we transcribe a media file, transcription entries will be added to the file. The original file is not altered but the information is added to it in the project file. Therefore, we need to switch off the read-only protection and put on the Edit Mode of a file. In the *Status Bar,* you can see whether a file is in Edit Mode or Read-Only. In the *Detail View* ribbon, a switch ☑ *Edit* allows you to switch on the Edit mode.

So before you start transcribing, your media file must be in Edit mode. Always check the status bar for *Editable* before you start your transcription. You will get a new **Edit** menu in the ribbon (largely the same icons as the **Audio** or **Video** menu).

Step 2. The Transcription Mode

In this example, we will show you how to transcribe the audio file of Helen. While transcribing in this example, we will refer in this example to the **Audio** menu. The procedure to transcribe a video file is identical to that of an audio

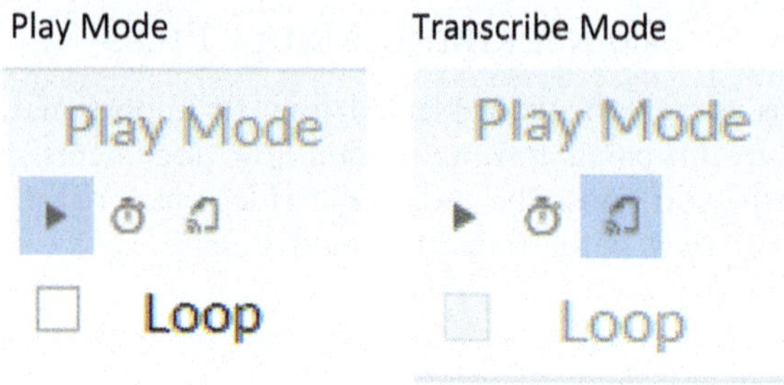

Fig. 6.8 Placing the program in transcribe mode

file. When you have a video file you want to transcribe, replace the **Audio** menu in this example with **Video**.

When opening a media file, the file will be opened in *Play* mode. That means that the Media Player buttons in the ribbon allow you to play, pause, and rewind the media file. To make a transcription, you must switch the *Play* mode to *Transcribe* mode in either the **Audio**, **Video** or the **Edit** menu (Fig. 6.8).

There is no immediate change visually, but be aware that the buttons of the Media Player (in the **Audio** or **Edit** menu) will now behave differently compared to what they did in Play mode once you activate the *Transcribe* mode.

Step 3. Make a First Transcript Entry

The transcription in NVivo is made in so-called Transcription Entries. These Transcription Entries are visible in the *Detail View* window on the right side of the Audio or Video. You can compare these entries with paragraphs in a textual transcription. As a user, you decide how much text is added in one entry. All entries are numbered, and NVivo automatically records the time stamp of each entry (Fig. 6.9).

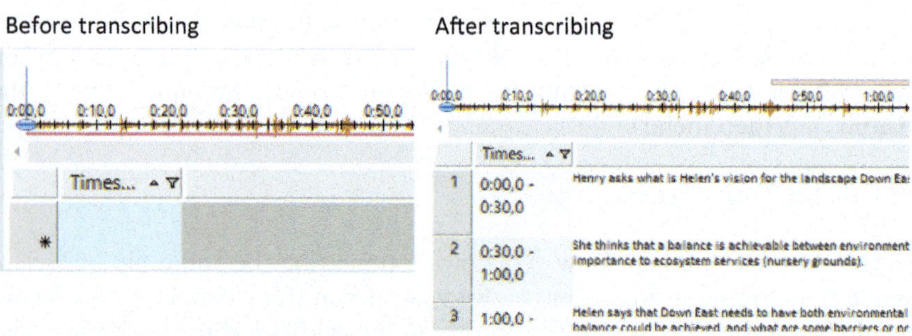

Fig. 6.9 Transcription entries in an audio file

To start your transcription, hit the *Play* button in the **Audio** or **Edit** menu. NVivo will automatically create a new Transcript Entry, allowing you to type the content there of what is being played. In the **Audio** or **Edit** menu, next to the play/pause and stop button, you can adjust both the Volume and the Play Speed of the audio. Adjusting the play speed is especially useful if you intend to transcribe Verbatim so you can more easily keep up with what is being said while transcribing. While playing, the *Play* button will act as a *Pause* button, so you can switch the audio on and off (do not press the *Stop* button at this stage).

Step 4. Repeat and Make Subsequent Transcript Entries

When hitting the *Stop* button, NVivo ends the transcript entry you are working on. When you use the *Play* button again, NVivo starts a new transcript entry. Subsequently, you can use Play and Pause alternating to determine the pace of your transcription. But each time you use the *Stop* button, a new entry is made. While transcribing, you might want to take advantage of the keyboard shortcuts of the media player: Play/Pause (F4), Stop (F8), Skip Back (F9) and Skip Forward (F10).

When you are finished, switch off the Edit Mode again in the *Detail View* ribbon, placing the file back in protected Read-Only mode. When you want to export your transcript to a file on your hard drive (either a Word file, PDF or plain text file): **RMC** (*above the name of the file*) > **Export > Export Video/Transcript…**. In the context menu *Export Options*, you can determine whether you want the media file to be exported, only the transcript or both. In the transcript, all three columns of the transcription window are exported: the numbers of the entries, the time stamps and the content.

Creating New Files

Transcribing media files is not the only way to create transcriptions of your primary material. You can also create a new document in your project and start typing a transcript while playing a media file with an external media player. This way, you can keep the project file size limited by leaving the media files outside of NVivo. The downside is that you do not have timestamps.

To create a new file, click on the folder where the file needs to be created. This is either in the Navigation View on **Data > Files** or one of your custom-made folders under **Files**. To make a new document, use **Create > Document**. Give the document a name, and the new document is created and opened in the *Detail View* window. As this is a new document, NVivo will automatically switch on the *Edit* mode so that you can start typing text (see also Fig. 6.7). Next to the **Document** menu, the **Edit** menu appears and provides text editing tools to determine the font type, the font size, the heading style, and layout options like bold, italic and underline.

When you are done creating the document, you can close the file with the ☒ at the right side of the file tab in the *Detail View* window. All document files can also be exported by **RMC** (*above the name of the file*) > **Export > Export Document ...**. Beware that, in this case, the time stamps are not saved to your external document. Since only text is created, only text will be exported.

CHAPTER 7

Working with Memos

Key messages in this chapter

- The literature only discusses memos. NVivo has two tools on board that fulfil the memoing function: memos and annotations.
- The Memo tool creates separate project items that function as independent documents in your project.
- The Annotation tool creates memos *inside* Files comparable to post-its to remind you of something important.
- A list of Memos and Annotations can be found in the section Organize under Notes.

Some Theory: The Use of Memos in Qualitative Research

Memos often have a mysterious status in qualitative research. For some, they are at the heart of the analysis and play a very comprehensive role (e.g. Charmaz, 2006; Richards, 2020). This role is negligible or not even discussed in others (Darlington & Scott, 2002; Patton, 2015). Memos are personal instruments that the qualitative researcher generally uses for themself and, in team research, also for fellow researchers. Glaser (1992) describes the memo as writing down ideas as they arise. These ideas can occur at any time during the research: when collecting the data, when writing the data or when (open)

D. Mortelmans, *Doing Qualitative Data Analysis with NVivo*, Springer Texts in Social Sciences, https://doi.org/10.1007/978-3-031-66014-6_7

coding. The idea behind the memo is that the researcher spends much time working on his material, and inevitably, ideas pop up in their head. They are often little more than flashes of thought that pass and disappear as quickly as they arise. The essence of a memo is that those flashes are important and should not be lost. They are the impetus for the development of the theory from the data. The researcher already makes connections between pieces of the data in their mind. Even though these thoughts are sometimes absurd and worthless, at other times, they are the start of a new insight that has emerged from the data and which the researcher can further develop. Remember that memo writing is given absolute priority over all other research work. When an idea comes to mind, drop everything and put the idea in a memo. This idea is based on the premise that insights do not necessarily have to come when the researcher thinks they should come. Valuable ideas, for example, are formed during coding or potentially even during non-research-related activities. Memos prevent these ideas from fading just because you did not plan for the idea to pop up at that moment.

Memos have a double relationship with the data the researcher collects. A memo is linked to data, and within NVivo, you can analyse a memo similarly to empirically gathered data (you can do essentially the same things to a memo as you can to a transcript during the analysis). So, a memo is treated as if it were data. Still, on the other hand, it is explicitly not stored in the folder *Data*, implying that it needs to be considered something different compared to empirically gathered data. It is a reflection of the researcher about the data. The researcher should try to preserve this double relationship when writing memos. NVivo partly solves this problem by allowing the researcher to write memos and link them to the data with a virtual link (if the researcher wants to). In this way, the connection remains without the memo becoming part of the data. When the memos are more theoretical, the researcher can clarify the link with the developing theory by, for example, using a keyword or giving the memo a title.

Memos or Annotations?

We distinguish different types of memos. When a researcher reads interpretively, theoretical ideas may arise depending on the ways of reading. We can use theoretical memos when the researcher records these in a memo. However, if he reads reflexively, thoughts may arise about how the data was collected, and he may be more likely to write *reflective* or *methodological memos*. But you may also comment on other team members (how they coded or built theory). These can be theoretical memos, but they can also be *commentary memos*.

The important role of memos in the analytic process is also translated into the software. NVivo provides two tools to keep track of memos, though the program only names one of them with the term "memo". In addition to "Memos", the program also has "Annotations", which essentially perform the

Table 7.1 Comparison of properties of memos and annotations in NVivo

	Memo	*Annotation*
Entity	Separate project item	Embedded in another project item and connected to a particular place in that project item
Coding	Coding the memo itself is possible	Coding the annotation itself is not possible
Export	Content can be exported to a file	Content cannot be exported
	A list of memos can be exported	A list of annotations can be exported
Storage in project	*Navigation view* > **Notes** > **Memos**	*Navigation view* > **Notes** > **Annotations**

same function as a memo but are used in a different place in the program with fewer functionalities and are therefore given a different name.

The NVivo tool with the name **Memo** is a short or longer text that is saved as a separate project item in your project. You can (optionally) link a memo to one (and only one) other project item. If you write a memo about a particular interview, it can be helpful to indicate to which interview that memo belongs. In that case, you will connect the memo to the project item (e.g. the interview transcript) using a memo link. In other cases, the memo will remain separate in your project as a memo document.

In addition to the Memo tool, NVivo also has a tool called **Annotations**. In NVivo, annotations take on the function of small scribbles in the margin of a document or information that is directly tied to a specific part of your data. For example, when the researcher is coding, he can take notes in the margin of a document near particular text passages. In essence, this is a memo, but NVivo offers the researcher a separate function for this: the annotation. If the researcher wishes, he can also create a memo and use it with a project item by linking it with a memo link. Conversely, this is impossible because an annotation is always attached *inside* a project item (usually a file with data), and annotations can never be created as independent project items. In short, an annotation has fewer possibilities regarding linkage (no memo link) or coding (cannot be coded in itself). Therefore, it could be seen as a "memo light", whereas the memo tool offers much broader analytical possibilities. In Table 7.1, we present a comparison between both tools.

Working with Memos

As the Memo tool creates a new project item, you can use **Create** > **Memo** to make memos. As a project item, you need to give the memo a name before it is opened (and saved in your project). A new tab opens in the *Detail View* window. As with documents (see Section "Creating New Files" in Chap. 6), a

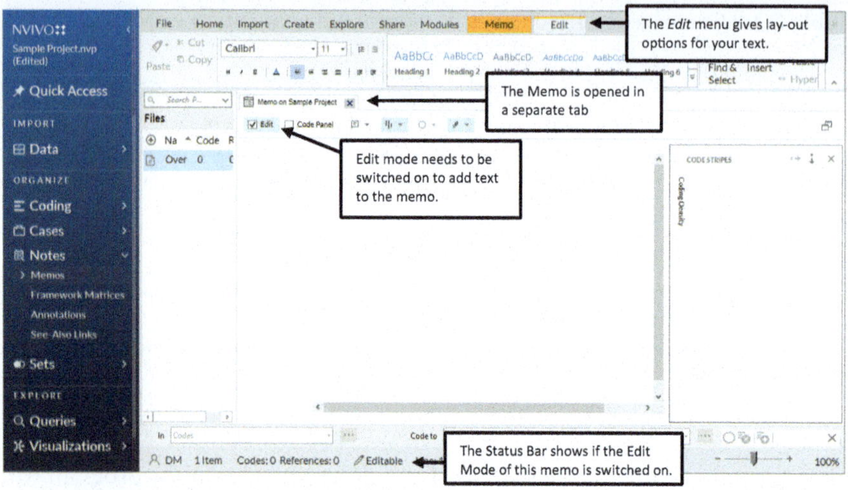

Fig. 7.1 Handling memos

new memo is immediately placed in Edit Mode, allowing you to start typing your memo. As shown in Fig. 7.1, an extra *Edit* menu with word processing tools is shown to change the layout of the text in your memo.

When you are finished, you switch off the Edit Mode again in the *Detail View* ribbon, placing the memo back in protected Read-Only mode. When you want to export your memo to a document, go to the list of memos in *Navigation View* > **Notes** > **Memos**, and then **RMC** (*above the name of the memo*) > **Export** > **Export Memo ….**

When you create a memo, the memo is saved in the project as a separate project item. There is no connection to any other element in your project. To connect your memo to other data pieces, NVivo allows you to link the memo with a Memo Link. Even though one can imagine that one would like to link a memo to multiple other objects (e.g. a memo on doctors' work practices linked to three interviews with doctors), there is only a 1-to-1 link possible. One memo can only be linked to one other project item. To link a memo to a project item, go to the item you want to link (e.g. a file in the section **Data**). Link the project item by **RMC** (*above the name of the file*) > **Memo Link** > **Link to Existing Memo …** and choose the name of the memo from the list that NVivo presents. It is also possible to create a linked memo from scratch by **RMC** (*above the file name*) > **Memo Link** > **Link to New Memo ….** When the Memo Link is established, you will see this with the icon in the File list (see Fig. 7.2).

To open the linked memo, **RMC** (*above the name of the file*) > **Memo Link** > **Open Linked Memo ….** Similarly, you can also delete the Memo Link.

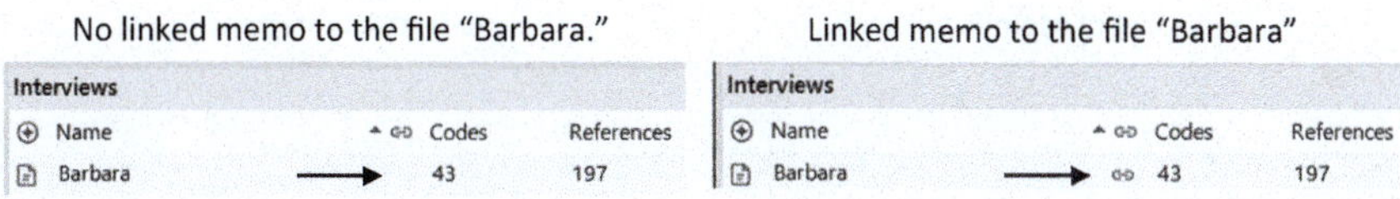

Fig. 7.2 Memo link in the File list

Working with Annotations

As the Annotation is a special kind of memo, the Annotation tool works slightly differently. An Annotation is meant for a (smaller) remark with a strong link to a specific part of your data, an observation you do not plan to incorporate into your coding in later stages of the research. Consequently, you need to open a file (whether textual, auditive, or visual) to create annotations.

When a file is open, an annotation is made to a selection of material in the file. For a document, this means you start with selecting a part of the text in the document. Next, you **RMC** (*above the selection*) > **New Annotation ...**. When the new annotation is created, a separate window opens at the bottom of your document where you can type the text of your annotation. Note that you do not get an extra Edit menu in this case, so a more advanced layout of annotations is not possible. You can only type straight text into the annotation field (see Fig. 7.3).

When you create an Annotation, NVivo will show a coloured background behind the selection to which the Annotation is attached. In the document's margin, NVivo shows an Annotation icon (▤) to indicate where an Annotation can be found.

An important consequence of the Annotation *not* being a project item is that the *Annotation* window itself does not have an ☒ to close the window.

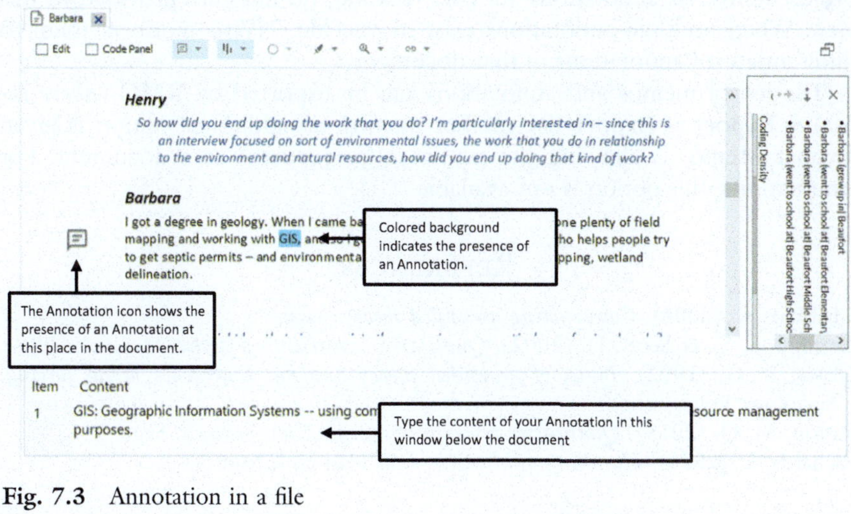

Fig. 7.3 Annotation in a file

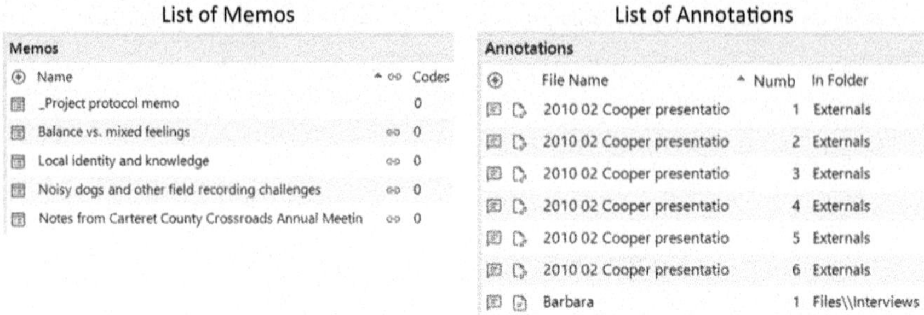

Fig. 7.4 Listing memos and annotations

You only see the ⊠ on the right side of the file tab in the *Detail View* window but not in the *Annotation* window. Nevertheless, you can close the *Annotation* window but not in a regular way by clicking on the ⊠ because that would close your document. To close the *Annotation* window only, go to the menu **Document > Annotations**. When the window is visible, the option *Annotations* will be selected, and you can close the *Annotation* window by unselecting the *Annotations* option there.

Listing Your Memos and Annotations

In the *Navigation View* > ***Notes,*** you can find a list of all your memos and annotations. In Fig. 7.4, the difference between both lists is illustrated. In the list of memos, you clearly see the self-chosen names of the memos, while for the annotations, only a reference is given to the document where the annotation can be found. Also, in the list of memos, you find a column with the icons of memo links, giving an overview of which memos have been linked to project items. As annotations are part of a file, no links are provided in this view. When multiple annotations exist in one file, NVivo numbers them to differentiate all annotations in that document.

The list of memos and annotations can be exported by **RMC** (*above the list*) > **Export > Export list ….** For memos, an additional option is given **Export Memo …** allowing the memo to be exported to a document. For Annotations, this option is not available.

References

Charmaz, K. (2006). *Constructing grounded theory*. Sage.

Darlington, Y., & Scott, D. (2002). *Qualitative research in practice*. McGraw-Hill.

Glaser, B. G. (1992). *Basics of grounded theory analysis: Emergence versus forcing*. Sociology Press.

Patton, M. Q. (2015). *Qualitative research and evaluation methods*. Sage.

Richards, L. (2020). *Handling qualitative data* (4th ed.). Sage.

CHAPTER 8

Thematic Coding

Key messages in this chapter

- Thematic coding is attaching labels to selections of data.
- NVivo foresees two ways of coding: by selection and by drag-and-drop.
- Codes are stored in a codebook, which can be organised hierarchically.
- Coding stripes and code highlighting are essential graphical tools to keep track of your coding process.
- Codes can be dynamically moved around the codebook.
- The aggregate function allows the researcher always to code close to the data and aggregate their coding work higher up in the code tree.

Some Theory: Coding in Qualitative Research

Data coding is one of the most important steps in a qualitative research project. The researcher can only arrive at reliable results in later analysis phases with adequately coded data. Coding means that the researcher *labels* all information that says something about the same subject and brings similar pieces of information together under the umbrella of a code. For example, if the interviewees describe a good relationship with the terms 'honesty' and 'reliability', ... then you can code this information under the general label of 'trust'.

D. Mortelmans, *Doing Qualitative Data Analysis with NVivo*, Springer Texts in Social Sciences, https://doi.org/10.1007/978-3-031-66014-6_8

Although the interviewees describe a good relationship in different terms, they essentially say the same thing, particularly that a good relationship must be based on 'trust'. Only by first labelling the data in a similar way can you do more complex analyses later. Coding is, therefore, the primary building block of a good analysis. In Grounded Theory, the coding of data is referred to as Open Coding. However, in the qualitative literature, many other terms can be found to describe the same process of labelling raw empirical material: categorising (Dey, 1993), labelling (Ritchie & Lewis, 2003), and initial coding (Charmaz, 2000).

In our example, the concept of "trust", which reflects the common factor behind the interviewees' answers, is called a code. You label or encode information on a specific 'code', and the collection of all these codes in a project is called a codebook. There are different approaches to coding in the qualitative literature. A first approach assumes that the codes are created before the coding of the data. These predefined codes are subsequently used for coding the empirical material. We call this the a priori or **deductive approach** (many other terms exist in the literature, though). A second approach is creating the codes while the researcher reviews the empirical material. That is the method followed, for example, in Grounded Theory. We will describe this method later as the **inductive approach**. The first way does not exclude the second. Even if you have already created a list of codes in advance, you can still add new codes during coding. So, coding strategies are often situated on the continuum between having all your codes predefined or creating codes while you go through the material.

What Are Codes?

Codes are labels. They are short terms or phrases given to a piece of data. The researcher reads through the transcripts or field notes and selects parts of the data (text, audio fragments or picture regions) relevant to the research. These selections are called references. One or more codes are then attached to such a reference, expressing the content of that reference (see an example in Fig. 8.1).

The researcher continues reading through the transcript and makes a new reference to which they attach one or more codes. In the reference in Fig. 8.2, two codes can be attached as two separate ideas pop up here: the high prices and the recent character of this problem. When the same ideas are expressed again (in the same interview or a different one), the researcher re-uses the code from the codebook that already fits this reoccurring idea.

The previous examples show the essence of coding: reading and labelling, reading and labelling. The codes must remain close to the data in this research phase (see also Section "Aggregating Coding Work"). In the initial coding phase (i.e. open coding), the researcher is not (yet) clustering the codes into concepts or themes. It comes down to enabling yourself to use your data in a structured manner later. The coding of your data is a kind of **data reduction**.

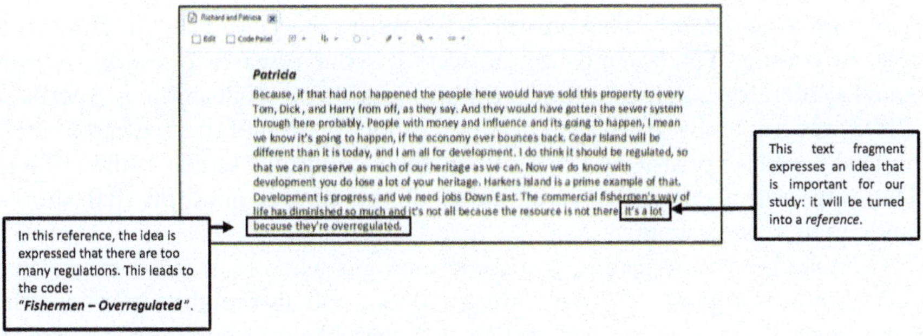

Fig. 8.1 Example of a code attached to a reference

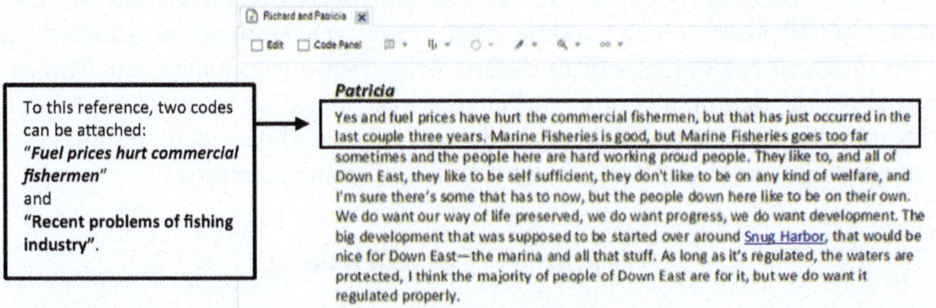

Fig. 8.2 Example of a code attached to a reference

Instead of reading all your transcripts repeatedly, you break down the whole transcript into smaller chunks of data (the references) and label these smaller pieces to get easier access.

These first examples of reading and labelling immediately elicit new questions:

1. What do I need to code?
2. How do I choose the name of a code? How do I determine the content of a reference?
3. How do I define the size of a reference?

These questions are dealt with in the following paragraphs.

What is Worth Coding in Your Data?

The first question we should ask ourselves during the coding process is: what content is relevant to be included in a code? This deceptively simple question is not easy to answer. After all, there is no reason to consider something relevant: relevance is a subjective concept. The most obvious answer is: "*It depends on your research question*". That answer is the most correct and the guiding principle in coding. You want to get answers to your research question in your analysis; therefore, that question must also guide you while coding. Auerbach (2003) recommends that researchers put a short version of their research question on a yellow post-it and hang it on their computer screen while coding. This way, you are constantly reminded of the central question that should guide your coding work.

Comparable to the feeling you sometimes get when reading literature that "everything is important", in a coding process, you also might experience the same "everything is important" feeling. You very likely end up in such a case by coding everything you see. At the start of your coding process, it is, therefore, best to have the coding work as closely as possible in line with your research question. The danger is that you exclude unexpected elements and surprises in the initial phase of your coding work. However, reading the same things from different respondents or in different observations can alert you. You can always revise the initial coding and change your mind about parts of the data you initially did not code. The insight that you might need to rethink your coding is also an example of an opportunity to make a memo.

Definition of the Reference

While coding, a researcher has to define a piece of data as a "reference". There are no rules here that apply to all situations in which you would want to code. For example, the researcher can determine how large the piece of data should be. In addition, he is not even bound by the fact that fragments must all be the same size. You can have both small and large references or overlapping references. But in general, three types of references often occur in qualitative coding work:

1. **Coding word by word (documents and websites)**

The most intense way of coding is going through the data word by word. The choice to code in such detail in some cases stems from the nature of the data one has. When you code official documents or internet pages, you can look at words and the meaning of words or the use of certain words in specific contexts in great detail.

2. **Coding by line or by paragraph (interview material)**

One sentence or a paragraph is the most commonly used reference while coding for textual material. You go step by step, line by line, through your data and see which ideas are contained in one line or paragraph. The reason for not taking too large a unit, like a long answer from a respondent for this process, has to do with the depth of coding discussed earlier. If your amount of text is too large, you, as a researcher, tend to use general codes more quickly. Going through your material line by line and seeing what it says brings you closer to the data and forces you to code more deeply.

3. **Coding event by event (observational material)**

Collecting rich observation data is a challenging task. When observing, the researcher must puncture the general social interaction and behaviour level. This difficulty persists in coding because observational material is difficult to code line-by-line. Therefore, an event is often taken here as a unit for coding. In this way, the researcher can compare different events with each other in their analyses and code them similarly or not.

Coding as a Mental Process

Coding is a process that involves the researcher going through an often complex mental process. After all, they must read the material and immediately assign a code (or even multiple codes) to a reference. Between reading and assigning lies a process of asking questions. The central question that the researcher asks himself is, "*What is this about?*". The second question is, "*Is this relevant to my research?*". Finally, the researcher proceeds to code: "*Which code should I attach to this?*".

In Fig. 8.3, we resume the previous example from the project file on the fishing industry but now explicitly address the mental process that the researcher goes through while trying to answer the research question: *why is the fishing industry in this town declining?*

Gaining Depth in Your Coding

This process is not "conscious", as shown in the previous paragraph. Moreover, the question we propose in this example is too simplistic. Strauss (1990) strongly advocates coding that theoretically digs as deep as possible from the start. After all, if you only ask, "*What does this mean?*" there is a significant risk that you will not get further than some general and descriptive categories. The purpose of coding is that your codes form the building blocks of the concepts we will develop in later phases. This means the researcher does not ask an overly simple question but a series of questions for each fragment. The researcher must be *critical* of their coding process and try to reach depth in their coding work. Charmaz (2005) gives some examples of more in-depth questions that the researcher should ask themselves when reading their data:

- What processes are going on here? How can I define these processes?
- How does this process develop?

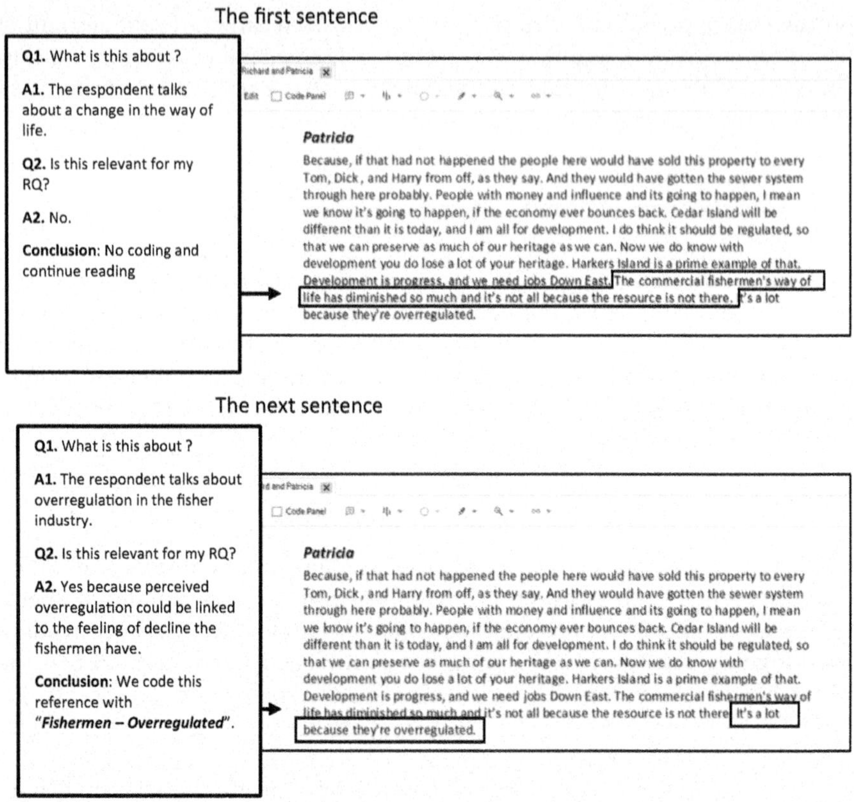

Fig. 8.3 Coding as a mental process of asking questions

- What do respondents do when the process occurs?
- What do respondents claim to be thinking or feeling during this process?
- What does the behaviour that I observe or that is described mean?
- How, when and why do specific processes change?
- What are the consequences of the process?

Glaser (1978) does not give general questions but instead proposes a sequence of questions that go from general to more specific:

1. Which research question does this reference answer?
2. What concept (or part of a theory) does this reference refer to?
3. What happens in this reference? What is the underlying process that the respondent has to deal with here?

These are just two examples of possible questions a researcher could ask. It's all about going beneath the surface. Codes can get stuck on the surface and be purely descriptive. If the researcher focuses too much on coding quickly (not asking many questions), the codes only enable you to grasp 'what's being said' (the thing about the fishing industry). Still, with superficial coding, your codes seldom help you understand why something is being said (why do people associate the industry's decline with...) or the necessary context needed to say certain things (only people from within a fishing family perceive...). Building your analysis on superficial coding often does not work very well during the later stages of the study.

Some authors state that codes should not have more than one word and a few words at most. This simplifies the labelling of coding, but for many researchers, this often leads to less refined coding, making the codes less useful later. One should never limit oneself to one word if that hinders the interpretability of a code later on. On the other hand, long sentences also do not work as "a label" any more and should be avoided. Glaser (1978) also advises including verbs in your codes to better preserve the dynamic character of social reality. Charmaz (2000) picks up on this and even states that the verbs in the codes help the researcher to see "through the eyes of the respondent" when coding. Active codes such as "influencing", "searching", and "learning" better reflect the respondent's experience than "influence", "search", or "knowledge".

Number of Codes and the Depth of Coding

The next question is how many codes can be attached to one reference. This partly depends on the size of your references (see Section "Definition of the Reference"). However, the number of codes attached to a reference also relates to the depth the researcher wants to achieve in their coding work. Some researchers create very general codes at one end and, therefore, need very few codes.

The code in Fig. 8.4 (Reasons) is so general that in further analysis, you will have to read and study many references with this code to gain an insight into the multiple reasons for the declining fishing industry.

However, the reverse is also possible: the coding in Fig. 8.5 is deceptively detailed. If you were to see these codes on a list, it becomes clear that this coding needs to be more abstract. Here, the researcher sticks so close to what the respondent says that the code doesn't really reduce the original reference in the transcript to its core idea with a 'label'. Coding is then reduced to selecting potentially relevant parts of the transcript without shortening the relevant parts to labels to keep everything manageable during the analysis. The researcher takes the regulations (potential cause) and the loss of jobs (possible consequence) together in one code, making the code so very specific that it's unlikely to reoccur in the same way in other data and thus creating once-only codes. If you have almost exclusively 'once-only codes', this might suggest you're being too specific in the labelling of codes.

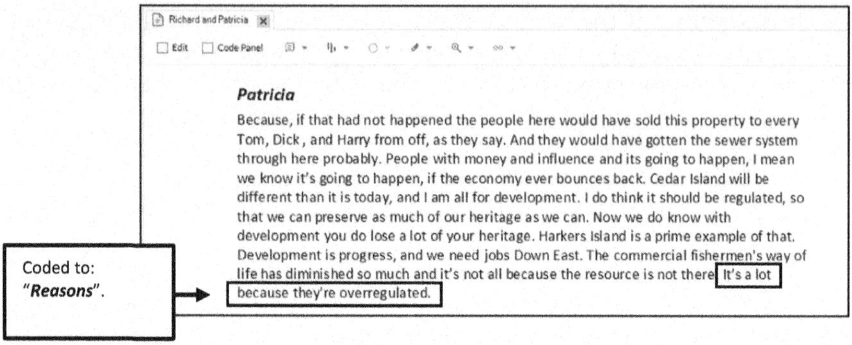

Fig. 8.4 Example of an overly general code attached to a reference

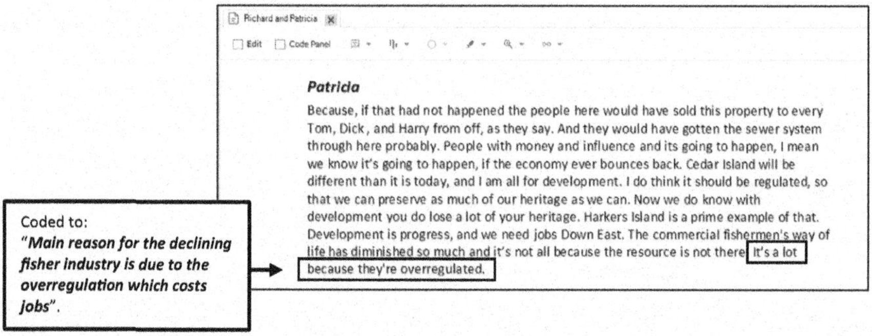

Fig. 8.5 Example of an overly detailed code attached to a reference

The number of codes attached to a reference also compromises the reference length (see further) and the coding depth. When you code on references consisting of one word, it is unlikely that you have ten codes attached. When you code to a paragraph in an answer, it could be the case that you have ten codes attached to this answer (provided that the respondents give much relevant information in this single paragraph).

Deductive Coding in NVivo

We presented earlier two ends of a continuum between starting your coding with a complete codebook on the one hand (deductive coding) and an inductive approach on the other. In this paragraph, we will explain how to work in cases with a codebook available at the start of your coding phase. We show the inductive coding approach in the next Section "Determining Your Unit of Analysis" in Chap. 9.[1]

This is the overview of the deductive coding process:

Prerequisites before you start

- Files with data imported in NVivo
- A codebook outside NVivo in Word, Excel or another QDA program.

Work plan

STEP 1. Create or import your codebook in NVivo
STEP 2. Code your files with this codebook

STEP 1. Create or import your codebook in NVivo

Looking at the menu **Import**, you will find an icon **Codebook** suggesting that NVivo foresees an import facility for codebooks. This import option *only* allows you to import Codebooks in the REFI-QDA format. REFI-QDA refers to the REFI-QDA Standard developed by the Rotterdam Exchange Format Initiative. This research group developed an exchange format for Qualitative Data Analysis software to export their codebooks or import codebooks from other packages. With this format, researchers can import codebooks from Atlas-ti or MAXQDA.[2]

Unfortunately, no conversion tool from Word or Excel codebooks to the REFI-QDA standard exists. So, for researchers working with a codebook outside another specialised QDA program, the import function will be of little help. In that case, The only alternative is to recreate the codebook in NVivo manually. New codes in the codebook are made through **Create > Code**. Each code is a separate project item, so you need to give the code a (unique) name. Next to the name, the field *description* might serve as a place to enter the definition of the code if the codebook foresees this.

When creating new codes in the codebook, you can immediately create a hierarchy in the codebook. NVivo allows for the creation of multiple layers in

[1] It is useful to read Section "Customising Your Workspace Outlook" in Chap. 5 on customizing your workspace to put your screen in Right hand view instead of Bottom View. During coding, Right hand view is usually a preferred layout of your workspace.

[2] For more information, see: https://www.qdasoftware.org/.

a codebook. Codes on a lower level are called "child codes", and the higher level codes are called "parent codes". NVivo will always close the hierarchy of a codebook. As shown in Fig. 8.6, when the hierarchy is closed, you can see at the + -signs where child codes are available. Since no ± sign is shown on the left of "parent code 2", you know this code has no lower-level children. Underneath "parent code 1" are two child codes visible when you click the ± sign.

When creating new codes with **Create > Code**, you must consider the starting position when you want new parent or child codes in your codebook. If you select the folder of the codebook in the *Navigation View* window, the **Create > Code** will add a top-level parent code to your codebook. When you choose an existing code in your codebook, the **Create > Code** will create a child code underneath the selected code. We illustrate this in Fig. 8.7.

In Fig. 8.7, we illustrate a two-level codebook. Of course, you do not need to limit yourself to these two levels. You can create a codebook with multiple layers, as shown in Fig. 8.8.

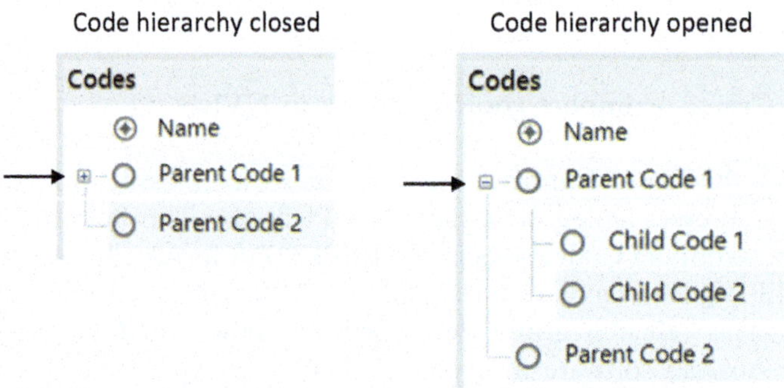

Fig. 8.6 Code hierarchy (open and closed)

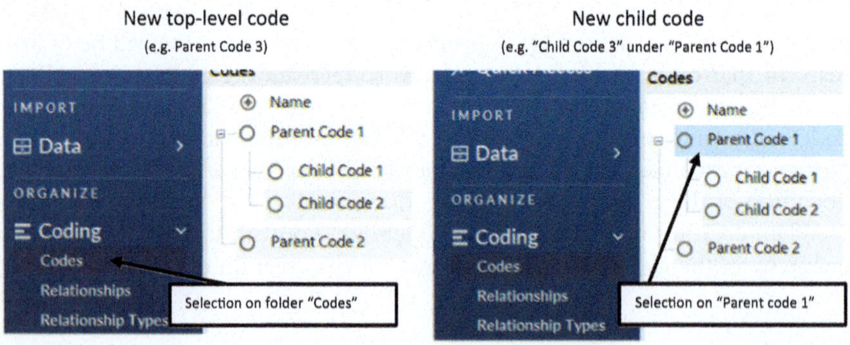

Fig. 8.7 Location of new codes, dependent on the selection in the codebook

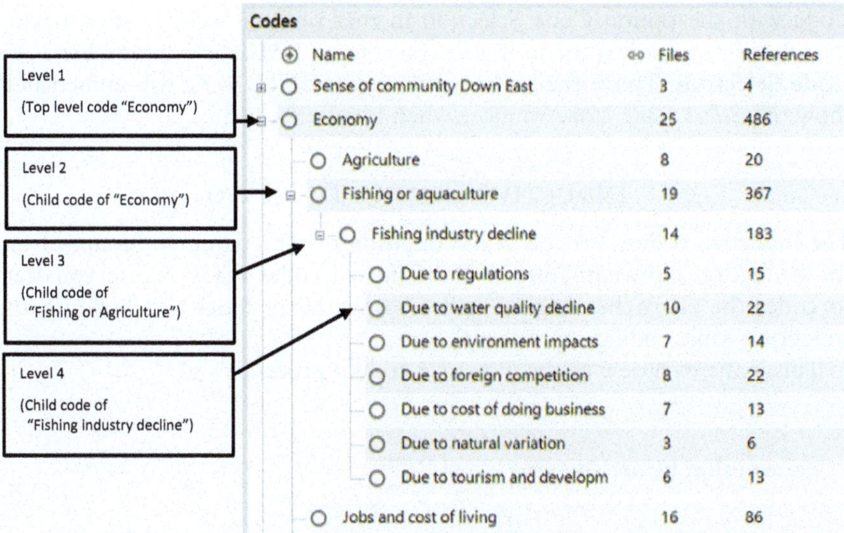

Fig. 8.8 Example of a four-level codebook structure

STEP 2. Code your files with the codebook

When the codebook is created in NVivo, the actual coding can start. Crucial to remember here is that codes are attached to references. In textual documents, coding *always* starts with selecting a text fragment (for other data types, we refer to Chap. 16). When a selection is made, you call for the *Select Code Items* window by **Document > Code > Code Selection ...**. In this window, you select all codes that apply to the selected text. The standard Windows selection operations apply in this window.

- Single click on code = Selection of one single code
- Click first code THEN Shift + Click on last code = Selection of multiple codes grouped together
- Press CTRL + click on multiple codes = Selection of multiple codes not grouped

When one or more codes have been selected in the *Select Code Items* window, the button at the bottom of the screen (*Code Selection* button) becomes blue. When clicking the button, the selected text reference is coded with all selected codes.

Repeat this process with the next text reference: selecting the text, selecting codes in the *Select Code Items* window and coding the reference with the selected codes.

To conclude this paragraph, it is worth noting that the *Select Code Items* window can also be called in three alternative ways. First, a small ribbon is available in the document tab of the file you are coding. There is also an icon

Code with the option Code Selection in that ribbon. Second, after making the selection, a context menu is also available by **RMC** (*above the selection*) > **Code Selection**. Third, the keyboard shortcut *CTRL* + *F2* will immediately show the *Select Code Items* window when hit after selecting.

Inductive Coding in NVivo

The inductive coding process is the opposite of the coding continuum from the deductive approach. Instead of having all codes ready before you start to code, the researcher begins with reading the material and building the codebook while coding the material.

This is the overview of the inductive coding process:

Prerequisites before you start

- Files with data imported in NVivo or created in your project file.

Work plan

STEP 1. Start reading your transcripts
STEP 2. Make new codes that apply to the data
STEP 3. Code the data with the new (and existing) codes

STEP 1. Start reading your transcripts

Since you have no codes to start from, the first step in the coding process is going through your material. To have a better overview of your material, it is recommended to read your material (e.g. the transcript you are about to code) as a whole without starting to code yet. When you first read the whole transcript, all details of the story of your respondent are refreshed in your memory. Next, when you start reading the text line by line and paragraph by paragraph to code the material, you still keep the overall picture of the story in your head, helping you get depth in the coding work.

The basic coding principle in NVivo is similar across all methods: all codes must be attached to a reference. To create a reference, you need to select part of your material. So, in an interview, a text fragment must be selected to enable coding.

STEP 2. Make new codes that apply to the data

When a part of the data is selected, you ask for the *Select Code Items* window by **Document > Code > Code Selection ...** (see the previous paragraph for alternative ways to get this window).

To create new codes, the *Select Code Items* window provides a button (on the right upper side of the window) to add codes to the codebook. As with the creation of new codes in the deductive coding procedure (see example in Fig. 8.7), the location of the new code is determined by the selection you make in the current codebook. When no codes are present in the codebook (the starting position), the new code will always be a top-level code. Once codes exist in your codebook and you add new ones, Fig. 8.9 shows the creation of new parent or child codes dependent on the selection in the *Select Code Items* window. When the folder of the codebook (e.g. Codes) is selected, a new top-level code is created. When another code is selected, a child code is created. The button to create new codes will also change its label (either "Top-level Code" or "Child Code") depending on your selection.

STEP 3. Code the data with (existing and/or) the new codes

When you create one or multiple new codes, coding the selection you made is identical to the deductive procedure: you select the codes that need to be attached to the reference, and you click the *Code Selection* button at the bottom of the *Select Code Items* window (Fig. 8.10).

In the inductive procedure, you will create many new codes at the start of the coding process. However, the purpose of the coding procedure is also data reduction, which implies that whenever an idea in your data is covered by a code already in your codebook, you re-use that code to code this reference. Consequently, as the coding phase progresses, fewer new codes will be

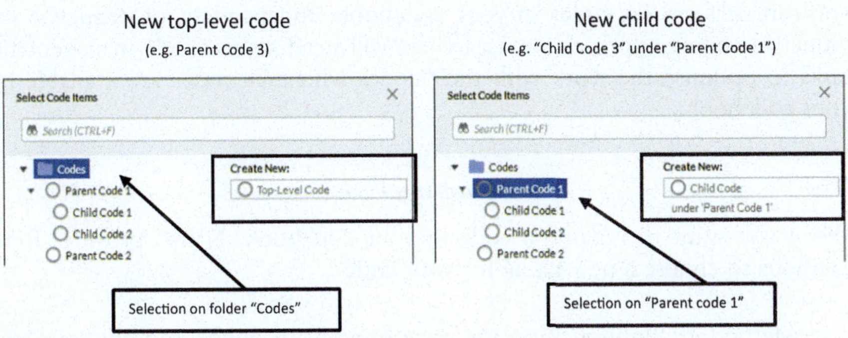

Fig. 8.9 Location of new codes, dependent on the selection in the Select Code Items window

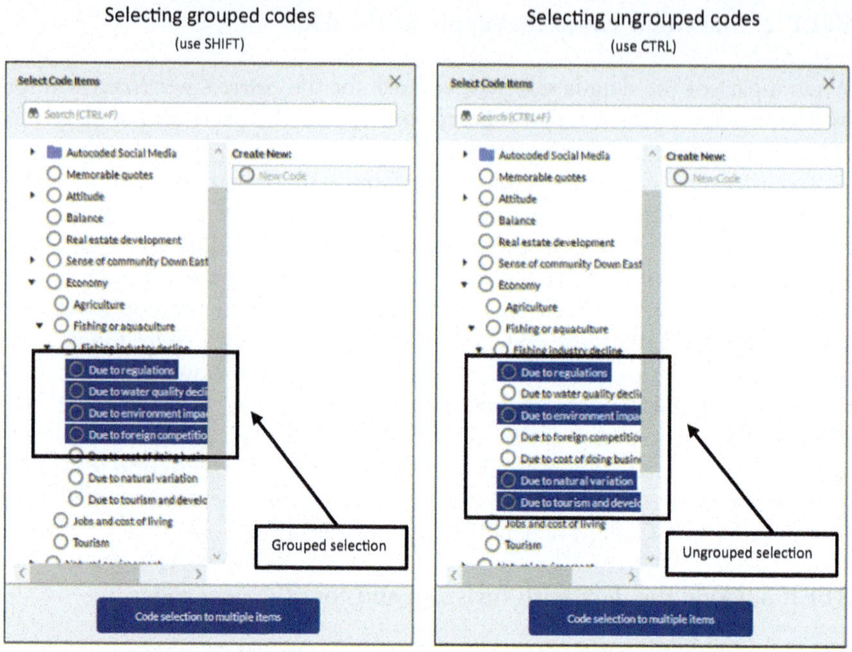

Fig. 8.10 Selecting multiple codes in the Select Code Items window

created, and at a certain point, you almost only choose existing codes from your codebook (much like in the deductive coding procedure).

Working with Codes in a Codebook

In the previous paragraphs, we concentrated on the primary task of coding, i.e. attaching codes to references of our data. This paragraph focuses on the tasks you can perform on codes in your codebook. Independent of deductive or inductive coding approaches, researchers will need several code-management tasks to optimise the work with their codes whenever codes are available in your codebook.

Renaming Codes

When you want to rename a code in your codebook, NVivo provides four methods to choose a new name for your code.

Method 1: Right-click on the code in your codebook and choose *Code Properties*. In the field *Name*, you can rename the code.
Method 2: Click on the name of the code and click **Home > Item > Code properties**

Method 3: Click on the name of the code and click a second time a bit to the side of where you clicked the first time. NVivo will make the name editable (the name gets a coloured background), and you can type the new name.
Method 4: Click on the name of the code and press F2 on your keyboard. The name of the code becomes editable in a similar way as in method 3.

Remark

- These methods also work identically when you want to rename a file or a case.

For the Mac user

- The Code properties can be found by **RMC** (above the name of the code) > **Get Info**.

Colouring Codes

NVivo gives the possibility to assign colours to codes. These colours can be added to codes by asking the properties of a code in **Home > Item > Code properties** or **RMC** (*above the name of the code*) > **Code Properties**.

A drop-down is available next to the option Color in the Properties window. When creating new codes, the standard colour is *None,* and the list of colours gives a wide range of colours to choose from.

Colouring codes can be used to give the Coding Stripes (see Section "Coding Stripes") a specific colour instead of the system colours chosen by NVivo. You can also use colours to identify codes used by specific users in your project or give particular parent codes (and their children) in the codebook a specific colour.

Remark

- Not only codes can be coloured. NVivo allows colouring of the following project items: files, codes, relationships, attribute values, and users. To give a project item a colour, always ask for the item's *Properties* to change the colour option.

Making Code Hierarchies in Your Codebook

Earlier (see Section "Deductive Coding in NVivo"), we already mentioned that the codebook in NVivo can be organised hierarchically with parent codes containing child codes underneath. We explained how hierarchies can be made by selecting either the code-folder (top-level parent codes) or other codes (child codes under the selected codes).

You do not need to create the hierarchy while you make new codes. Often, researchers coding inductively do not know beforehand what their coding tree will eventually look like. Consequently, they usually start by creating a series of codes on the top level, and after coding a few files, they re-organise (or, better, hierarchise) their codebook.

The first method to create hierarchies is to use the Windows shortcuts *CTRL + X* (cut) and *CTRL + V* (paste). When you want to move a code to another place, you select the code and cut the code (*CTRL + X*). Then, you choose where the code needs to go and paste the code (*CTRL + V*). You need to consider the selection protocols that we explained in Fig. 8.7.

The second method is the drag-and-drop method. You can drag codes around in your codebook and drop them anywhere to create or undo a hierarchy. Parent codes can be turned into child codes by dragging them to another code, and child codes can be turned into top-level codes by dragging them to the folder **Codes** in the *Navigation View*. This is illustrated in Fig. 8.11.

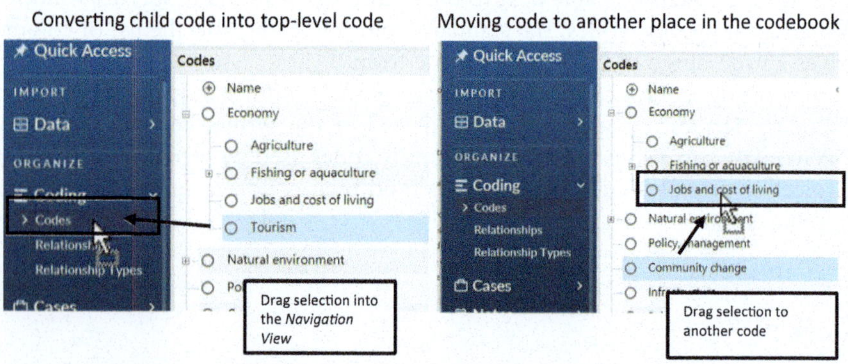

Fig. 8.11 Organizing hierarchies in the codebook

Remark

- The drag-and-drop in NVivo works slightly differently from the regular drag-and-drop in Windows (e.g., Explorer). In Windows, you click on an item, and you immediately can start dragging the item. In NVivo, this is a two-step operation. First, you select the item by clicking on it, and second, you click on the selected item again and start dragging.

Determining Your Sort Order

When you create codes, NVivo stores all codes in the section **Codes**. When codes accumulate, you get a codebook that you can organise hierarchically (see previous paragraph). NVivo will sort this codebook alphabetically. This is visualised in the list of codes at the column head. Next to *Name*, a black triangle (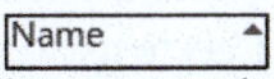

) shows that the sort order is placed on the name of the codes. The pyramid shape shows that the sort order goes from A to Z. Clicking on the column changes the sort order: the triangle becomes a reversed pyramid, indicating sorting is done now from Z to A. You can click on any of the other columns of the *List View* of your codebook to have this column determine the sort order. For example, clicking on the column head of *References* will sort the codebook from the code with the lowest to the highest number of references.

Some researchers want complete freedom to determine the order of their codes in the codebook as they wish (for content or easy accessibility of the first codes in the tree). This can be achieved by custom sorting the codebook. As custom sorting is not available by default, you must switch from automatic to custom sorting. This is done in **Home > Workspace > Sort by > Custom** after selecting a code in the codebook. Once the custom sorting is activated, you can **RMC** (above the name of a code) > ***Move Up*** or **RMC** (above the name of a code) > ***Move Down***. When you select multiple codes, moving a group of codes up or down in the custom sort order is possible. If you have many codes, use the keyboard shortcut CTRL + Shift + U to move codes upwards or CTRL + Shift + D to push them down.

Efficiency in Coding

In this paragraph, we discuss some tools NVivo offers to help researchers with their coding work and, more specifically, to gain efficiency while coding.

Drag-and-Drop-Coding

In Section "Deductive Coding in NVivo", we explained how to code references with the *Select Code Items* window. The idea behind this window is to give the researcher an overview of all codes available in the codebook and to let the researcher select the codes that apply to this reference. All selected codes are then attached to the reference *at the same time*.

A second way to code is known as drag-and-drop coding. To use this method, you first activate the *Code Panel* in the detail view window of the file you want to code. An option *Code Panel* can be checked in the document tab's small ribbon. On the right-hand side of the document, the Code Panel with an overview of all codes in the codebook is shown.

To code a reference, start with selecting the text fragment. Next, you click on the selected text and drag the text to the Code Panel. When your mouse enters the Code Panel area, a code gets selected, and an arrow-icon (↗) appears at the bottom of your mouse pointer (see Fig. 8.12). Move your pointer to the code that needs to be attached to this reference. When you release the mouse, the reference is coded to this code.

You can keep the reference selected and drag the reference a second time to another code. That way, the reference is coded under a second code as well. Repeat this until all codes that apply to this reference have been attached to the reference. When you have a large codebook, you might want to put the code you want to use "in sight" by moving the slider to the right of the *Code Panel* up or down.

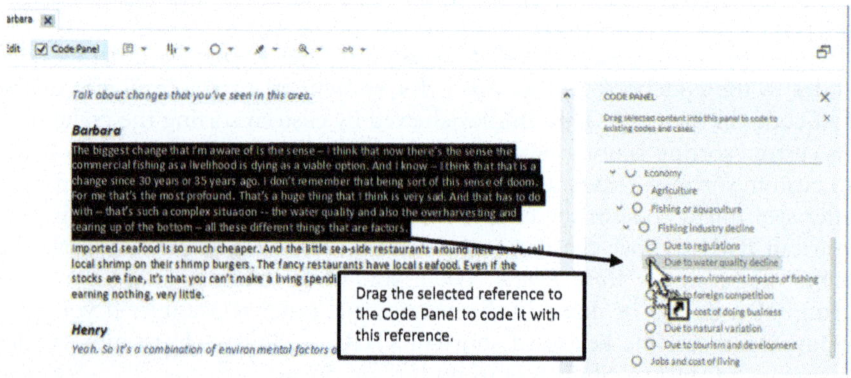

Fig. 8.12 Drag-and-drop coding with the Code Panel

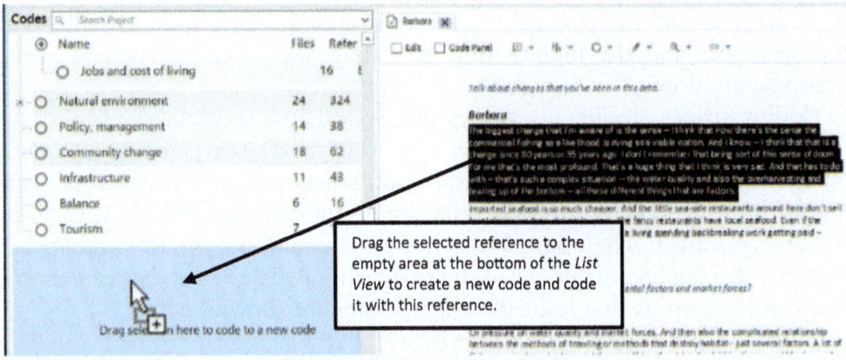

Fig. 8.13 Drag-and-drop coding with the codebook in the list view (create a new code)

> **Remark**
>
> - Many researchers find the drag-and-drop coding a quick and efficient way of coding. However, the classic method using the Select Code Items window is more efficient in certain situations. Specifically in cases where you have an extended codebook and want to attach multiple codes to a reference. Learn to work with both methods and decide which is more efficient for your work.

You can use the coding panel to do drag-and-drop coding, but alternatively, you can also use a drag-and-drop style coding directly to the codebook itself. When you open a file in the *Detail View* window and put your codebook in the *List View*, you can also drag your selected references to the codebook in the *List View*. An additional feature of this method is that NVivo shows a white area at the bottom of the *List View* where you can drag references that need a new code. When dragging a reference to this bottom white area, NVivo will ask you to give the new code a name. The program will subsequently create the new code in your codebook and code the reference with that new code. Instead of the arrow-icon (⧉), your mouse pointer will now show a plus-icon (⊞) (see Fig. 8.13).

Coding Stripes

In the previous examples, we showed how you can code your data. However, when you code, the program has no immediate feedback on whether the coding was successful or what codes have been attached to which reference. In this paragraph and the next one, we present two tools that give some direct visual feedback to the researcher about the coding work done in a file. The coding stripes and the highlighting of coding allow you to visualise better what you are doing.

Coding stripes are literally stripes that appear in a separate window at the right-hand side of a document that is opened in the *Detail View* window. Only opened files can show coding stripes, so you must open a file first before using this feature. The coding stripes of a file are default switched off and need to be activated manually. The options to activate the coding stripes are presented in Document > Coding Stripes or in the small ribbon. The researcher can choose between six options that regulate which stripes are shown:

1.	All	Coding stripes for all codes are shown (up to the maximum stripes defined in the options (usually 7). You can increase the number of stripes up to 200 (not recommended)
2.	Selected items	Coding stripes are shown for a user-defined selection of codes
3.	Coding density only	Only the Code Density (most left stripe) is shown and no additional stripes are offered for specific codes. The higher the density, the darker the coding density stripe will be. A white colour signifies the absence of coding at this place in the file
4.	Most coding	Shows only the codes that have been used the most in this file
5.	Least coding	Shows only the codes that have been used the least in this file
6.	Recently coded	Shows only the codes that have been used most recently during your coding work

When you are coding, the option *Most recently coding* is recommended as this gives you a check of whether your coding is going as expected. When your coding is done, the options *Most* and *Least* can shed light on the most or least dominant themes. The *Selected Items* option is useful when for example splitting codes (see Section "Splitting Codes").

A crucial element in the use of coding stripes is their colour. Using different colours per stripe makes the difference between the codes easier to spot. NVivo offers two ways of getting your coding stripes coloured: *item colours* and *automatic colours*. The default value is *item colours* whereby NVivo uses the colours the researchers gave to the codes (see Section "Colouring Codes"). The disadvantage of this default option is that when researchers have not given any colours to their codes, all coding strips are white and no differences are visible between the stripes. In that case, changing the colour scheme to Automatic colours in Document > Coding Stripes or in the small ribbon is better. Figure 8.11 illustrates the difference between the different possibilities (Fig. 8.14).

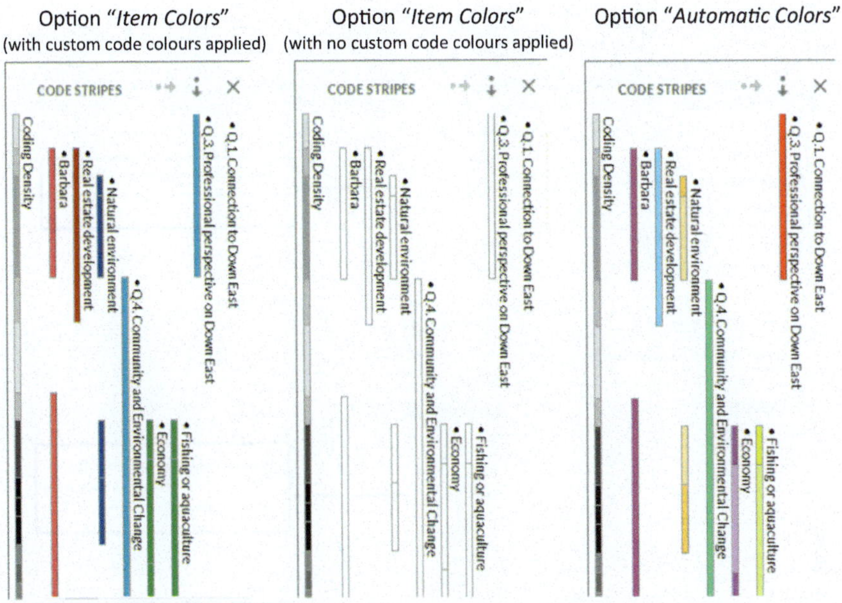

Fig. 8.14 Different effects of colour schemes in the coding stripes

> **Remark**
>
> - The Coding stripes are not only shown in the *Detail View* window of a File. They are also available in the *Results* window of queries.

Code Highlighting

The second tool to visualise your coding work is code highlighting. Code highlighting foresees a coloured background (usually yellow) behind your data when codes have been attached to that section of the data. Visually, you can see the differences between yellow and white areas as the coded and not-coded areas of your data. As with the coding stripes, the code highlighting is also switched off by default. You can switch it on by **Document > Highlight** or in the small ribbon under the same icon. Two options are available: *All Coding* or *Coding for Selected Items*. These options work the same way as the similar options of the coding stripes (see previous paragraph). When you only work with thematic coding, the option *All codes* is the most useful as it shows your thematic coding work. When you also work with Case coding (see Chap. 9), it might be wise to use *Coding for selected items* and only select the codes in your codebook and not your cases to avoid everything being coloured due to the coding with cases (see Section "Making and Using Case Classifications" in Chap. 9).

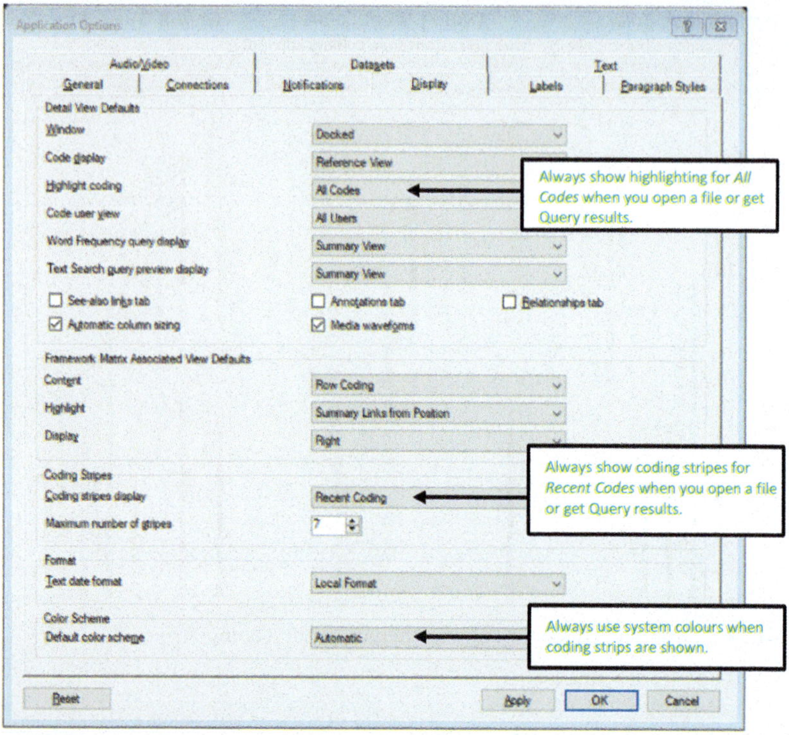

Fig. 8.15 Options window, tab display

Remark

- In the Windows version of NVivo, you can permanently turn on the Coding Stripes and Highlighting. Whenever you open a file or get Query results, the Coding Strips or the Highlighting (or both) will be shown.
- Go to File > Options and choose the tab Display to run these options. Here, you can adapt the default behaviour for Coding Stripes and Highlighting and also select your colour scheme (Fig. 8.15):

Code Highlighting with Coding Stripes

Two crucial visualisation tools were shown in the previous sections: coding stripes and code highlighting. Coding stripes are shown in a separate window on the right-hand side of your coded data and represent coded sections of your data. The code highlighting foresees a yellow background behind that coded data. NVivo foresees a combination of both when you select one particular coding stripe.

When Coding Stripes are switched on, you can click on each coding stripe separately. The effect of selecting one coding stripe is that the data that is coded with that code is highlighted with the same colour as that of the coding stripe. In the example in Fig. 8.16, we select the purple coding stripe “Economy”. Consequently, the coded text in the interview of Barbara is highlighted with a purple colour. Clicking on a different coding stripe changes the highlighting and the colour of the highlight.

In the small ribbon, you can see that the options of coding stripes have been changed as well. When we activated our coding stripes, the option *All* was selected. After selecting a particular coding stripe, the option automatically changes to *Custom*. To return to the previous situation, you need to re-select *All*.

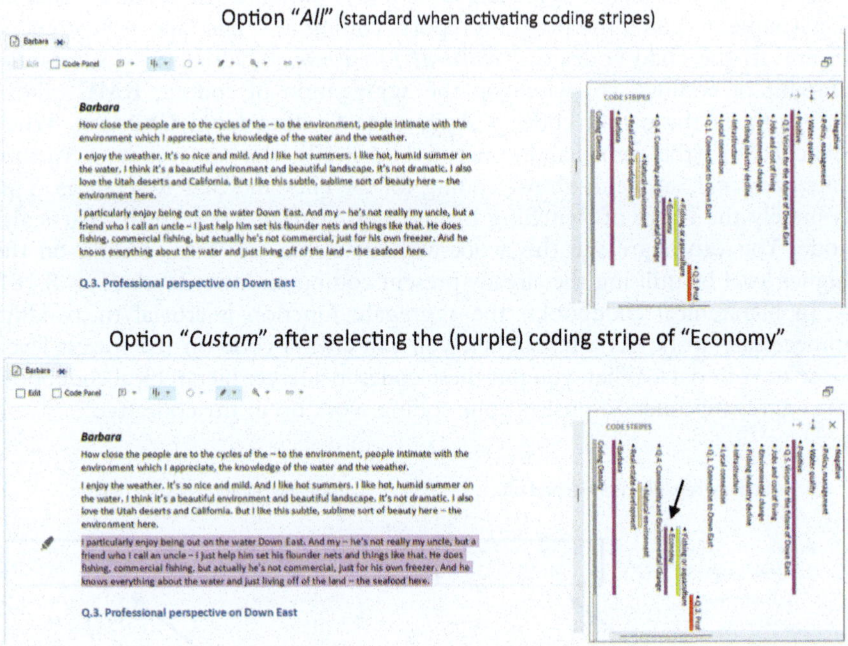

Fig. 8.16 Effect of selecting one Coding Stripe (activation of custom coloured highlighting)

Some Final Coding Tips

To end this chapter, we present a few extra tips for coding in NVivo. They can help researchers in specific situations. Most of these tips help you gain efficiency in your coding work or subsequent analysis.

Aggregating Coding Work

Previously (see Section "Number of Codes and the Depth of Coding"), we suggested always coding *as close to your data as possible*. Coding is labelling the content of your data, and the codes need to express the content of what your respondents have communicated. After coding the data, you will organise your codebook and turn it into a hierarchical codebook. Higher-level codes in your codebook (e.g. see Fig. 8.8) are often more abstract than the lower-level codes. To avoid duplication of coding work at the higher levels, NVivo allows the user to aggregate the coding work from the lower-level child codes to the higher-level parent codes. The effect of aggregating coding is shown in Fig. 8.18. When the aggregate function is switched off for the "Natural Environment" code, the program reports coding in 0 Files and 0 References. However, the child codes of "*Natural Environment*" do show a substantial amount of coding. To switch on the aggregation of coding, **RMC** (above the name of the parent code) > **Aggregate Coding from Children**. When switched on (the right-hand panel in Fig. 8.17), the same code, "*Natural Environment*", suddenly shows coding for 24 Files and 324 References. This is merely the effect of switching on the aggregate function in this particular code. You can aggregate this work without doing any extra coding on the higher level by utilising the already present coding on lower levels (Fig. 8.18).

In hierarchical codebooks, the aggregate function is crucial to avoiding unnecessary work. As a researcher, you can always code on the lowest level, close to your data. When you put these codes in a larger hierarchical codebook, the aggregate function pushes your coding work up in the code tree.

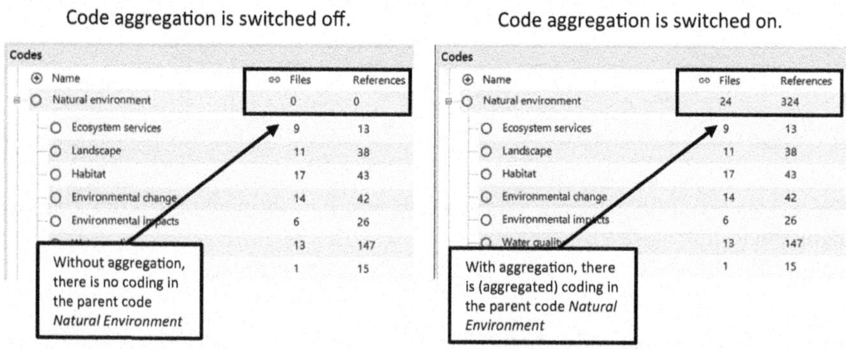

Fig. 8.17 Code aggregation switched off (left panel) or on (right panel)

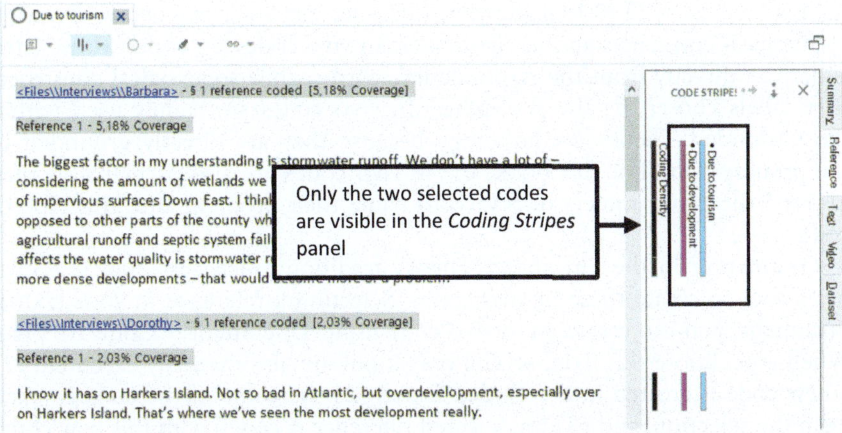

Fig. 8.18 Limiting the coding stripes view with selected items …

> **Remark**
>
> - The aggregate function pushes coding work *one* level up (and strictly only 1 level). Take, for example, the four-level code structure in Fig. 8.8. The code "*Economy*" is the level 1 code. When you activate aggregation in this code, only the coding work from level 2 (codes "*Agriculture*", "*Fishing or aquaculture*", and"*Jobs and cost of living*") are aggregated. The codes at the level-3 code, "*Fishing industry decline*", are not aggregated to the level-1 of "*Economy*". To aggregate the coding from the level-3 code "*Fishing industry decline*" all the way to the level-1 code "*Economy*", you need to activate the aggregate function both in the level-1 code "*Economy*" (aggregation from level 2 to 1) and in the level-2 code "*Fishing or aquaculture*" (aggregation from level 3 to 2).
> - A quick way to aggregate all coding work across your codebook is to select all codes in your codebook (CTRL + A) and then **RMC** (above the selection) > **Aggregate Coding from Children**. This way, the aggregation is applied to all codes in your codebook and the coding work is aggregated from the lowest to the highest levels in your code tree.

In Vivo Coding

A specific type of code that a researcher can use is the In Vivo code. The term In Vivo code comes from the Grounded Theory methodology. In his classic textbook, Strauss (1990) made the difference between two types of codes: sociological constructs and in vivo codes. In vivo codes are codes whose label is derived directly from the respondents' words, while sociological constructs have labels chosen by the researcher. In vivo codes are mainly used within the Grounded Theory methodology because they are directly grounded in the primary data. For this book, the in vivo codes are also particular as they inspired the programmers of NVivo to name their program after this type of code.

Originating in the Grounded Theory tradition (Richards, 2002), NVivo has a tool to create these in vivo codes. Selecting a reference in your textual document, you can create an in vivo code using **Document > Code In Vivo**. When you choose this icon, several operations are automatically executed: (1) a new code is created in the **Codes** folder, (2) the name of the code is taken from the selection, and (3) the selected reference is coded with this new code.

A few things to watch out for when using this function. First, the new code is always placed in the folder **Codes** even if you have subfolders with different codebooks. If the new In Vivo code must be placed in one of these folders, manually drag them to the correct folder. Second, the separate icon *Code* In Vivo might suggest a different type of code next to the regular codes you create while you code manually. This is not the case. The new code made when you select the icon *Code* In Vivo is a regular code and is indistinguishable from other non-in-vivo codes. Only the label will remind you of the in vivo character of this code. Third, NVivo will automatically take the *whole* selection, meaning the entire selected reference, as the label of the code. Therefore, the in vivo coding function only works when you select one or a small number of words as the name of your code. Avoid selecting a whole paragraph if you want to use this function because the label of the new code will be non-sensical or unpractically long.

Uncoding

When you code your material, you will code many references with a substantial amount of codes. During this process, you will make mistakes or regret some coding decisions. In this case, you want to *uncode* or detach the code from a particular reference. NVivo offers two methods to uncode references: quick decoding and elaborate decoding.

Method 1: Quick decoding of one code

The first method to decode a code from a reference requires activating your coding stripes. Decoding a code can then be done by **RMC (above the coding stripe) > Uncode**. Behind the RMC of a coding stripe, there is always the option to uncode, making this the quickest way to decode a single code from a reference if your coding stripes are active.

Method 2: Decoding multiple codes from a reference

Sometimes, multiple codes have been attached to a reference, and you want more control over which codes need to be decoded from that particular reference. To selectively decode multiple codes from a reference, select the reference that needs uncoding. Then call for the *Select Project Items* window by **RMC**(above the selected reference) **> Uncode**. In the *Select Project Items* window you can now choose the codes that need to be removed from the reference. Be aware that you are reversed-selecting now: what you select will be *un*coded (unlike the coding mechanism where you selected codes to be coded to a reference). In the *Select Project Items* window, you will see that all codes (and relationships and cases alike) that are attached to this reference are shown in bold black. Only these items can be selected to be uncoded. The light grey ones have never been attached to this reference and can therefore not be chosen for uncoding.

Merging Codes

Coding is manual and labour-intensive work. You need to see data in your references and label the content. In addition, you need to classify the same content under the same codes to reduce data with the coding. In large projects, it is very plausible that researchers lose track of some codes and create new codes for ideas that other codes have already captured. When going through a codebook, one might find such "double" codes expressing the same ideas under different labels. For such cases, NVivo allows merging codes so that one of these "double" codes disappears and the coded references merge into the other code.

To merge two codes, first, you must decide which code will disappear and which will receive the coding work from that first code. The merge starts with the code that will disappear. Locate the code in the codebook and **RMC > Cut**. At this moment, the code is given a faded colour. It seems half-removed but its content is still available in the memory. Next, go to the code that will receive the coding work from the first code and **RMC > Merge Into Selected Code ….** The first code disappears from the codebook and its content is merged into the second code.

Remark

- You can achieve the same result with aggregation of codes. When you want to merge two codes, you can make a new parent code and move both codes underneath this parent code. When you aggregate the coding from these two child codes (see Section

"Aggregating Coding Work") you also have a code that contains the coding work of both codes but without removing one of the original codes.

Splitting Codes

Splitting codes is the reversed operation from merging. In this case, the coding work is too general to be of use. E.g. see the example in Fig. 8.4 where a general code "Reasons" was used. Such a code is too broad to be of service in the further stages of your analysis. One solution for the use of codes that are too general is to recode your material again. Another solution (only applicable to a few of these codes) is splitting the coding work between two or more codes. Unlike the merging of codes, there is no built-in function in NVivo that allows the easy splitting of coding work. In this paragraph, we present a workaround to achieve the redistribution of references from one to two codes. The example can be extended to more than two codes if applicable.

In the sample project, you will find a code called "*Due to tourism and development*". This code can be found under the parent code "*Fishing industry decline*" and contains references where respondents refer to the cause of the decline in tourism or development. For example, we now want to split this code into two codes: "*Due to tourism*" and "*Due to development*". To achieve this goal, you proceed in four steps:

Step 1. Duplicate the code

Go to the codebook and copy the code that needs to be split (*CTRL+C*). Immediately paste the code into your codebook (*CTRL+V*). You now have a duplicate code in your codebook. Since NVivo requires unique codes, you will see that the "(2)" was added after the name of the original code. So in your codebook, you now have the following two codes:

- *Due to tourism and development*
- *Due to tourism and development (2)*

Step 2. Renaming both codes to their new label

The second step is to rename these two codes so that they are labelled as you want them after the splits are performed (for renaming techniques, see Section "Renaming Codes"):

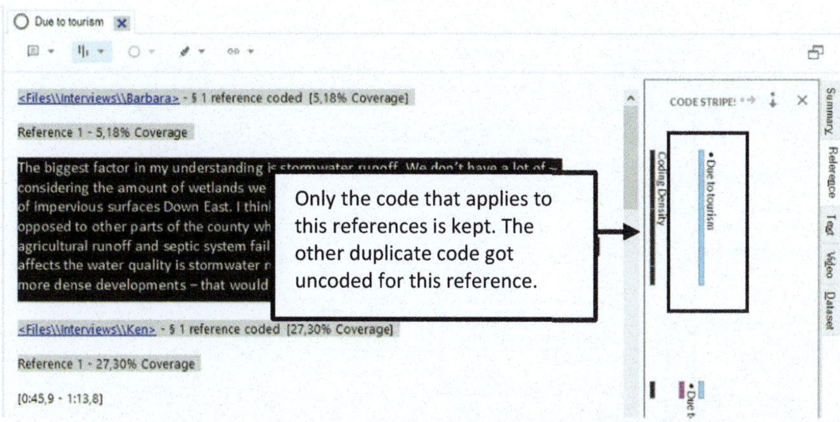

Fig. 8.19 *Uncoding at the coding stripes to split a code into* two codes

- *Due to tourism*
- *Due to development*

Step 3. Open the first code and adapt your coding stripes

When you double-click on the first code, NVivo opens a new tab with all the references of that code. You activate your coding strips and choose *Selected items* .... In the window *Select Project Items* you now select only the two codes you want to split: "*Due to tourism*" and "*Due to development*". All other codes need to be deselected. As a consequence, you will see only two coding strips now next to the references:

Step 4. Uncode the code that does not apply to this reference

In step 3, all Coding Stripes are showing double now. That is the consequence of duplicating the original code. Both codes have identical references coded to both codes. To split the original code in two, you can easily uncode the code that does not fit with this reference by **RMC** (above the coding stripe that needs uncoding) > **Uncode** (Fig. 8.19).

Printing Documents with Your Coding Stripes

NVivo can print your documents with the coding stripes (to the printer or a PDF document) if you want to review your coding work on paper. First, open a file in the Detail View window to get text and coding stripes together

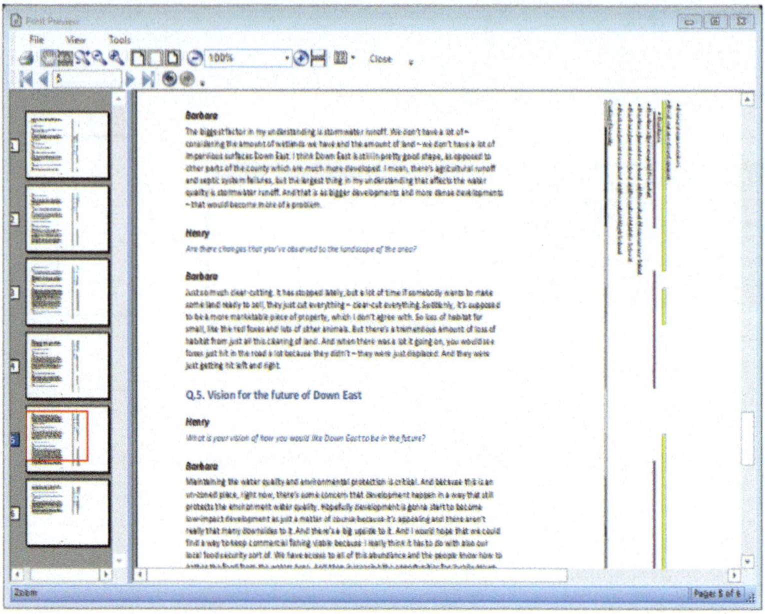

Fig. 8.20 Print preview window showing *an interview text and the coding stripes*

on a page. Next, ensure your coding stripes are activated in the *Detail View* window (see Section "Coding Stripes" for more information). Go to **Share > Print > Print Preview** to get the *Print Options* window. In that window, the option *Coding Stripes* already suggests "*Print on Same Page*". You can also have the coding stripes printed on adjacent pages if you want. Click OK to get the *Print Preview* window (see Fig. 8.20) from which you can send the document to a printer or a PDF document.

References

Auerbach, C. F., & Silverstein, L. B. (2003). Qualitative data. An introduction to coding and analysis. New York University Press.

Charmaz, K. (2000). Grounded theory: objectivist and constructivist methods. In N. K. Denzin & Y. S. Lincoln (Eds.), *Handbook of qualitative research* (2nd ed., pp. 509–535). Sage.

Charmaz, K. (2005). Grounded theory and the 21st century: Applications for advancing social justice studies. In N. K. Denzin & Y. S. Lincoln (Eds.), *The Sage handbook of qualitative research* (3rd ed., pp. 507–536). Sage.

Dey, I. (1993). *Qualitative data analysis*. Routledge.

Glaser, B. G. (1978). *Theoretical sensitivity: Advances in the methodology of grounded theory*. Sociology Press.

Richards, T. (2002). An intellectual history of NUD*IST and NVivo. *International Journal of Social Research Methodology*, *5*(3), 199–214. https://doi.org/10.1080/13645570210146267

Ritchie, J., & Lewis, J. (Eds.). (2003). *Qualitative research practice*. Sage.
Strauss, A. L. (1990). *Qualitative analysis for social scientists*. Cambridge University Press.

CHAPTER 9

Coding with Classifications

Key messages in this chapter

- Coding with classifications is a different type of coding than thematic coding.
- NVivo offers two classification tools: File Classifications and Case Classifications.
- Classifications are tools to make comparisons of descriptive data (e.g. socio-demographic background) easier.
- File classifications are meant to be used in Literature reviews. Case classifications are used to compare background characteristics of people or organisations.
- Importing a descriptive matrix from Excel is the most efficient way of working with classifications.

Some Theory: Comparing Basic Operation in Qualitative Research

Constant comparing is a central activity in the Grounded Theory approach (often called the constant comparative method) and in all qualitative analysis methods. Results can only be shaped when the researcher recognises patterns in their data. Those patterns can only come to the surface when they code in the same way elements of meaning that are the same (see the previous chapter on thematic coding). Because you work with qualitative material, this

D. Mortelmans, *Doing Qualitative Data Analysis with NVivo*, Springer Texts in Social Sciences, https://doi.org/10.1007/978-3-031-66014-6_9

is a subjective interpretation of the researcher. In quantitative research, you ask a question, "How do you feel about this?" and the respondent can answer with predefined categories such as "good", "neutral" or "bad". In qualitative research, each respondent has their way of telling and explaining how they feel. One respondent said straightforwardly: "I don't feel good about that at all". Another respondent will phrase this more subtly: "That is difficult". Still, other respondents may disguise their answer as "Let me give an example...".

Comparisons are made differently, and the researcher will constantly search for the most meaningful comparisons. A first way of comparing is by comparing themes in your thematic codes. In the sample project, part of the story respondents tell describes the fishing industry's decline. The researcher will use codes in the codebook (e.g. "*due to regulations*" or "*due to water quality decline*") to compare different reasons given by respondents about the phenomenon at hand (the decline of the fishing industry). But this way of comparing ignores differences between respondents. Often, qualitative researchers are also interested to know whether the content of the interviews also differs across types of respondents. Therefore, a second way of comparing consists of determining whether the reasons for the decline in the fishing industry vary across gender or labour market status. The difference in this type of comparison is that the information for the comparison is not coming from the transcript but rather from the properties of the transcript (i.e. the metadata). The comparisons are done with data that describe the material: "This interview was done with a woman" or "This interview was from a commercial fisherman".

Performing this second type of comparison in NVivo is precisely the topic of this chapter. NVivo uses classifications to make comparisons between groups of respondents easier. Before we explain how this type of comparison is integrated into the program, we first show in Section "Out-of-the-Box: What If There Were No Classifications?" how these types of comparisons would be done *without* any classifications.

Out-of-the-Box: What If There Were No Classifications?

To illustrate the power of cases and classifications to make comparisons in NVivo, we start by showing how these comparisons would be made if these tools were not available in the program. With the information from Chap. 8, all comparisons that we will enable with cases and classifications can also be performed with regular thematic codes in our codebook.

Let's assume we want to make three comparisons in the sample project: across gender, age groups and educational level status. To enable these comparisons, we first include three parent codes in our codebook and add the comparison categories below as child codes (see Fig. 9.1):

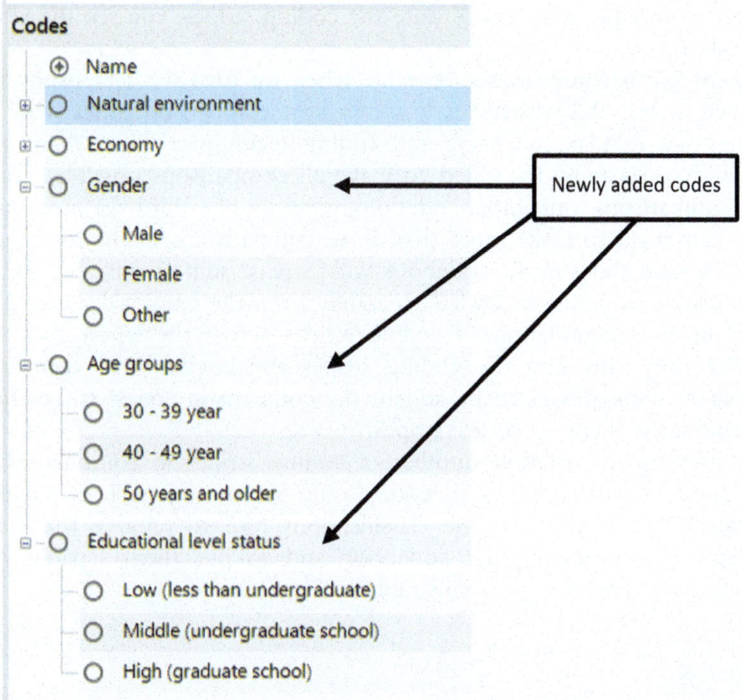

Fig. 9.1 Codebook with comparison codes gender, age groups and educational level status

In the original codebook of the sample project, we had 36 codes. We added another 12 codes, or an increase of one-third of the initial codebook, for the three comparisons we want to make.

The next step is to code our material with the comparison codes (see Chap. 8 for thematic coding techniques). As an example, we will code the interview of Barbara. We chose the child codes for this respondent: woman, 40–49 years old and middle (undergraduate school). Since the interview with Barbara is collected in a fixed moment, her gender, age and educational status do not change across the data collection of this single interview (which can be different, for example, in observational field notes). Therefore, we code the whole interview of Barbara with these three child codes. To code an entire interview, we select the file of Barbara in the folder **Files > Interviews**. To code the entire interview, **RMC** (above the file name) > **Code Whole Document**. NVivo will warn, "Are you sure you want to code the whole document?" Click on *Confirm* to get the *Select Code Items* window (see also Section "Deductive Coding in NVivo" in Chap 8). Now select the three child codes "*woman*", "*40–49 years*", and "*Middle (undergraduate school)*" (use CTRL on your keyboard to select the ungrouped codes) and click on the button below: *Code Selection to multiple items*. When you open the interview by double-clicking, and you activate the coding stripes, you see the effect of this coding.

Figure 9.2 becomes better readable when we turn the screenshot 90° as is shown in Fig. 9.3 which clearly shows how comparison codes work when you use the regular codebook and coding techniques. First, many parent and child codes must be added to make all comparisons possible. For only three comparisons, our initial codebook grew by one-third. In most studies, researchers want to make more than three comparisons, so the total number of comparison codes in the codebook will increase substantially. Second, each comparison code must be coded separately from the empirical material. This results in an accumulation of coding stripes across the whole file. Again, Fig. 9.3 only illustrates the coding stripes panel with three comparisons. With more comparisons to be added, the comparison codes' stripes quickly overwhelm the thematic codes' stripes.

To prevent an irritating number of coding stripes in combination with a codebook filled with lots of codes, only meant to enable a handful of comparisons, NVivo offers the classifications tool to remove the comparison codes from the regular codebook and to put them separately in a "classification-database" as is illustrated in Fig. 9.4.

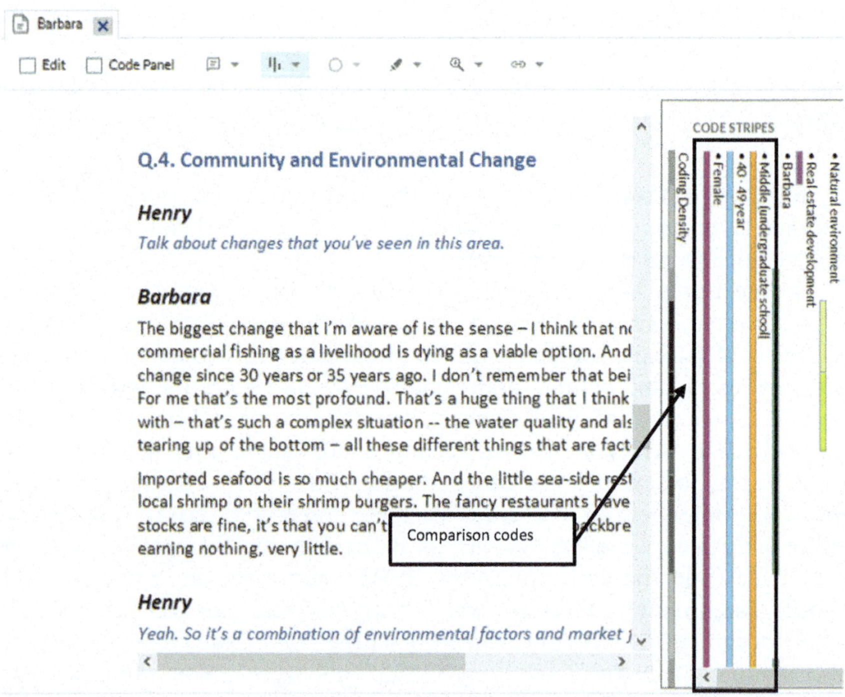

Fig. 9.2 Interview with coding stripes showing the thematic codes and comparison codes

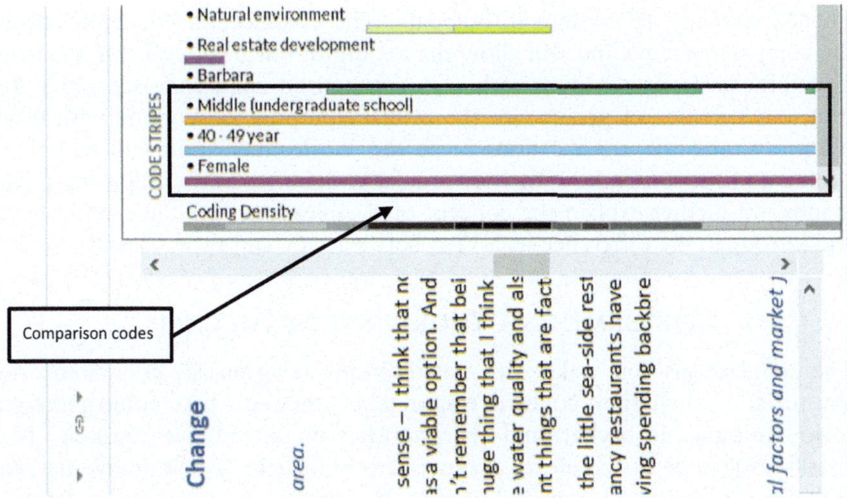

Fig. 9.3 Rotated interview with coding stripes showing the thematic codes and comparison codes

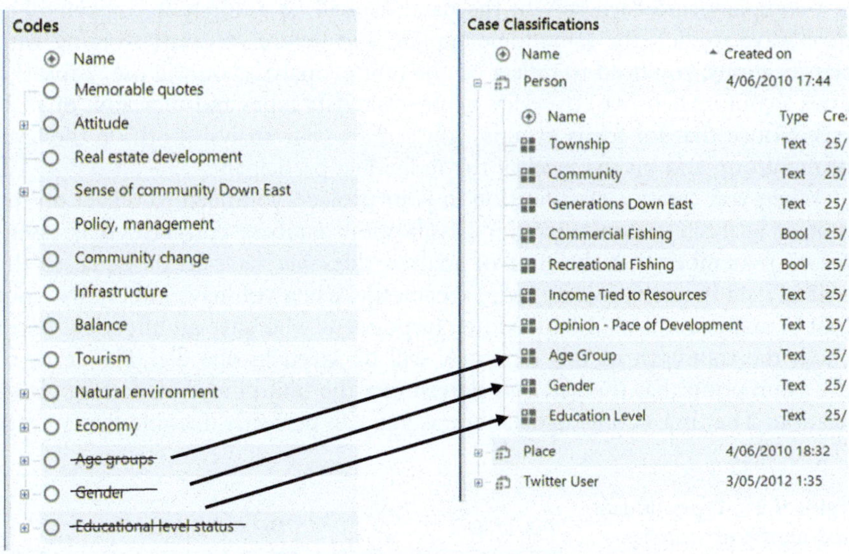

Fig. 9.4 Moving comparison codes to attributes in a case classification

Classifications can be considered as a separate database in your project. Inside the database, NVivo stores variables (called *Attributes*) that allow to make comparisons. The child codes we had to create in our codebook are now categories (called *Values*) of the variables in the database. In this way, our codebook can be restricted to merely the thematic codes that identify labelled elements of content in our data. The classification efficiently stores the comparative data and will allow the coding of our data with this database more efficiently than coding with each comparison child code separately. In sum, comparisons of groups are the primary purpose of working with classifications, and the classifications make this work much more efficient than adding comparison codes into your regular codebook. In the following paragraphs, we further explain the concept of classifications, attributes and values and introduce the difference in NVivo between File and Case Classifications.

Determining Your Unit of Analysis

The comparisons you make with classifications are primarily comparative in nature, and when they concern respondents, they are best comparable to socio-demographic background characteristics in quantitative research. The classification tool is, therefore, also perfectly similar to a quantitative dataset with variables in the columns and cases in the rows. As in a quantitative dataset, the rows determine the unit of analysis. If your survey has interviewed humans, the unit of analysis will often be individuals, and each row will represent one individual. When your survey concerns economic entities, the unit of analysis is a company, and each row in the database will be a different company. For qualitative studies, the same logic applies, and before you start working with classifications, you need to reflect on the unit of analysis you will use. Table 9.1 gives a (non-exhaustive) overview of possible data types and their unit of analysis. Notice that for focus groups, you can do your analysis both on the level of the group and on the level of the individual participant.

When you create a classification in your project, you need to reflect on the unit of analysis to construct the classification database correctly. It is essential to remember that the unit of analysis does *not* have to be equal to the unit of data in your project. More specifically, when you have interviews, your unit of analysis will be the individual respondent who gave an interview. Most likely, the transcript of this interview will be saved in one file. This leads to the assumption that this *file level* determines the unit of analysis for the classification. That is a wrong association, as you can perfectly imagine that we still

Table 9.1 Types of data and unit(s) of analysis

Type of data	*Unity of analysis*
Interviews	Respondent/Individual
Focus groups	Focus group
	Participant/Individual
Case studies	Case
Media messages	A newspaper article, A tweet, ...
Literature	A scientific article
Law data	A ruling from a judge in a case

have multiple files for one respondent in our project: the interview transcript, a methodological report, a picture of a piece of paper on which we asked the respondents to draw their household relations, etc. So, when you reflect on the unit of analysis, do not reflect in terms of project items (files) in your project but in terms of what determines the type of comparisons you will make. So, when doing interview research, your unit of analysis is the individual and not the transcript of one interview.

Choosing the Correct Type of Classification

Even though the classification tool is programmed in a relatively uniform way in the program, NVivo distinguished between two types of classifications: file classifications and case classifications. When we look at the pre-defined classification schemes that NVivo offers, it shows how the programmers of NVivo think about the difference between these two types of classifications (see Fig. 9.5).

The list of pre-defined File Classifications shows predominantly Literature types like *Electronic Book* or *Encyclopedia,* while the list in the Case Classifications is limited to *Organization* and *Person.* Even though we also find *Interview* or *Focus Group* in the list of predefined File Classifications, the message is clear that File Classifications are meant to be used in literature reviews. On the other hand, case Classifications are used for the more common data types in qualitative research: interviews with persons, case studies in organisations, focus groups, etc. Even though you are free to use File Classifications in these types of studies as well, we advise you always to use Case Classifications unless you use NVivo for Literature reviews (in that case, you do not have much of an option; see Chap. 21).

Table 9.2 also shows that some studies will require more than one classification in their project. For example, in a focus group study, you likely will make a case classification *Person* for analyses where the unit of analysis is the participant in the focus groups and a second case classification *Focus Group*

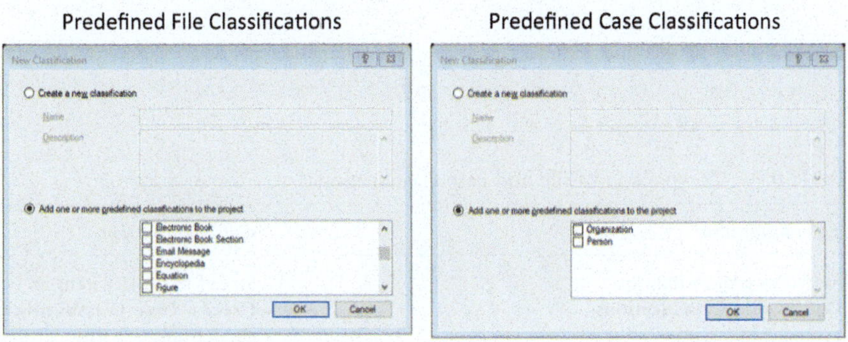

Fig. 9.5 Predefined file and case classifications

Table 9.2 Types of data, unit(s) of analysis and recommended classification types

Type of data	*Unity of analysis*	*Classification type*
Interviews	Respondent/Individual	Case classification
Focus groups	Focus group	Case classification
	Participant/Individual	Case classification
Case studies	Case	Case classification
Media messages	A newspaper article, A tweet, ...	Case classification
Literature	A scientific article	File classification
Law data	A ruling from a judge in a case	Case classification

for comparisons where the unit of analysis is the focus group as a whole. We further elaborate on focus group analysis in Chap. 15.

The unit of analysis determines the content of the classification, but more is needed to explain the technical differences between File and Case Classifications thoroughly. In Table 9.3, we compare both tools.

While Table 9.3 focuses on the differences, both tools have many similarities regarding the construction and definition of the instrument. Both File and Case Classifications are databases with Attributes as their main components (comparable to variables in a quantitative dataset). Each Attribute can have Values (similar to categories in a categorical variable) that define the differences between groups of data. When starting to work with classifications, the definition of a classification differs from the construction of a (quantitative) database. Figure 9.6 shows the graphical parallels between a classification in NVivo and a quantitative database.

In the following paragraphs, we explain (1) how to create the classification instrument and (2) how to code your data with the tool. We first discuss the File Classification (Making and Using File Classifications), followed by the File Classification (Making and using Case Classifications).

Table 9.3 Comparison of file and case classifications

File classification	*Case classification*
Stored as project item in **Data > File Classifications**	Stored as project item in **Cases > Case Classifications**
Coding inside a file	Coding inside a case
Not applicable in the coding query	Applicable in all queries

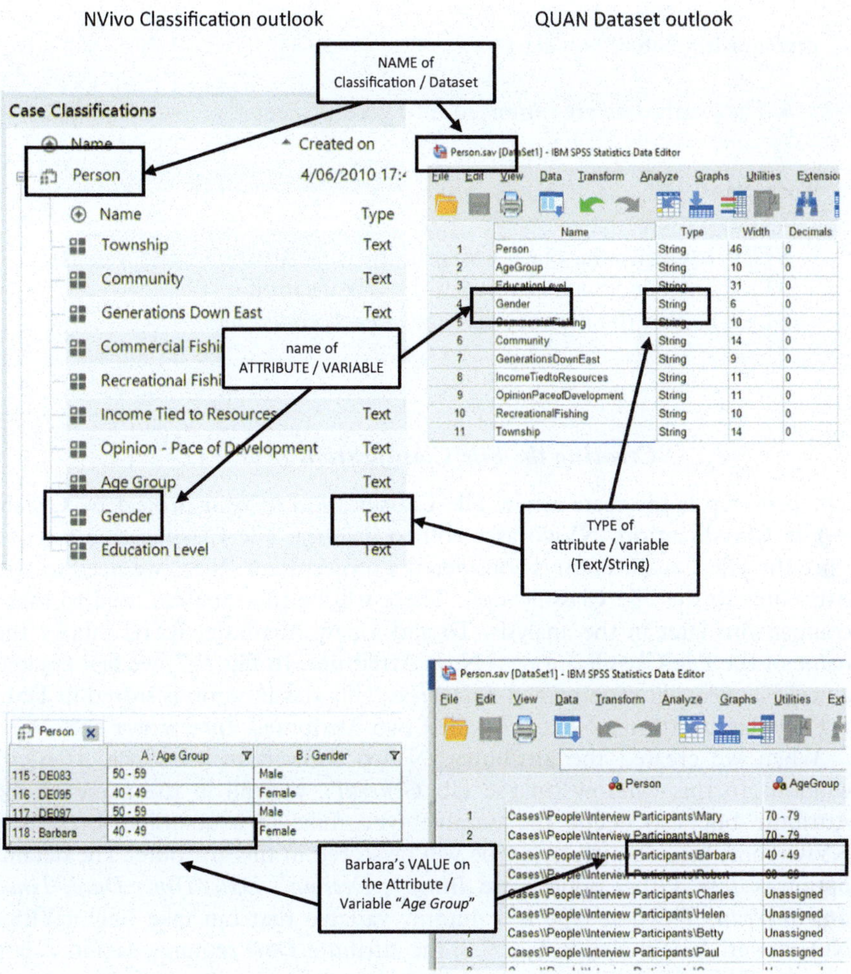

Fig. 9.6 Comparison of NVivo classification and quantitative dataset

Making and Using File Classifications

Even though File Classifications are mainly used in Literature Reviews, we start with explaining the File Classification as creating and using this tool requires fewer steps. Therefore, a new user should read this paragraph before learning how to construct the more commonly used Case Classification.

This is the overview of how you work with File Classifications:

Prerequisites before you start

- Files with data are imported into your project.

Work plan

STEP 1. Create the File Classification (with Attributes and Values)
STEP 2. Code your File(s) with the classification
STEP 3. Add the file-specific data to each coded file.

Creating the File Classification (STEP 1)

The first step is to create a new File Classification in your project by **Create > File Classification**. NVivo asks you to give the File Classification a name since the File Classification is considered a project item. Next, we need to add Attributes to the File Classification. These will be the variables used to make comparisons later in the analysis. To add a new Attribute, **RMC** (above the name of the File Classification) > **New Attribute**. In Fig. 9.7, we first created a new File Classification named *Interview*. This classification is stored in **Data > File Classifications**. Next, we added two Attributes: *Interviewer* and *Date*.

When we created the attributes, NVivo showed us the *New Attribute* window. In this window (in the tab *General*), you fill in the name of the Attribute but also define the variable type. Like in a quantitative dataset, NVivo must know which data type will be stored in this attribute. The default option is *Text*. Other options are *Integer*, *Decimal*, *Date/Time*, *Date*, *Time*, *and Boolean*. This last type is a dummy variable that can take only a YES/ NO answer. As is shown in Fig. 9.8, the attribute *Date* requires data in a date format. Therefore, we changed the default type option *Text* to *Date*. Now, only dates can be entered in this Attribute. For attributes with the type *Text*, we can define specific values that this attribute can take.

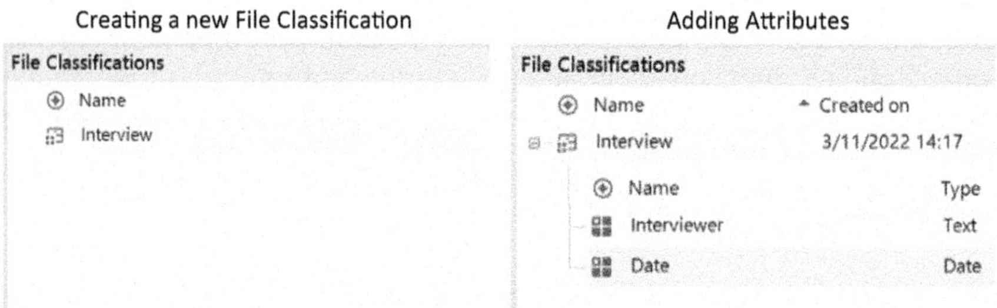

Fig. 9.7 Creating a new file classification (left panel) and adding attributes (right panel)

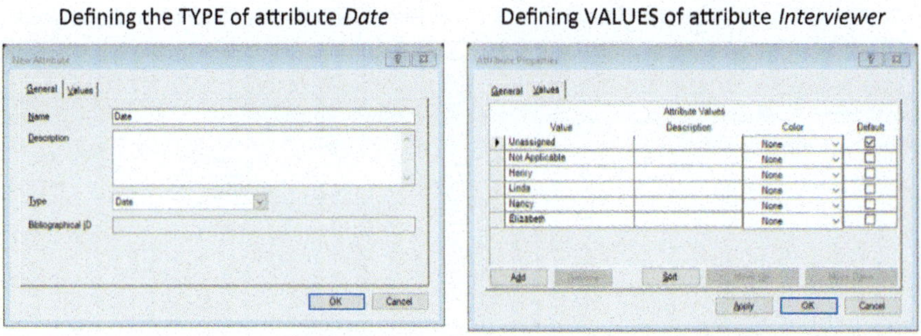

Fig. 9.8 Defining type of attribute (left panel) and values (right panel)

Defining Values (in the tab *Values* of the *New Attribute* window) will help you later to enter the data in your project consistently, as all values you define here will be shown in a drop-down menu when entering the file-specific data. In the right panel of Fig. 9.8, we show the values entered for the attribute *Interviewer*. With the *Add* button, we added four values to this attribute: Henry, Linda, Nancy, and Elizabeth. NVivo will always define two values for a Text attribute: Unassigned and Not Applicable. These two cannot be removed nor renamed. The value *Unassigned* (which refers to a missing value) is always used as the default value. When new data is entered into the project, NVivo will always fill in *Unassigned* until the researchers enter their own data. Keeping the default value always on Unassigned is safe to avoid incorrect data entry later.

For each comparison you want to make later in your analysis, you must define a separate attribute. Remember that you always have the opportunity to add attributes later in your analysis. So, you do not need to be exhaustive when creating the classification.

Using the File Classification (STEPs 2 and 3)

Once you have created a File Classification with some attributes, you can code your files with the classification (STEP 2) and fill in the file-specific data (STEP 3). Since these two steps are done in the same window (see further), we discuss both steps together in this paragraph.

In the previous step, a File Classification was created with two attributes. This is merely an empty database now without any connection to the data in our project. In step 2, we will code our material with the classification to use the instrument in our project. To code a file with the File Classification, you **RMC** (above the name of the file) > **Document properties**. You go from the tab *General* to the tab *Attribute Values* (see Fig. 9.9). When the file is not coded, you see *No Classification* in the option *File Classification*. To code your file with the file classification, use the drop-down to select the file classification that you want to add to this file.

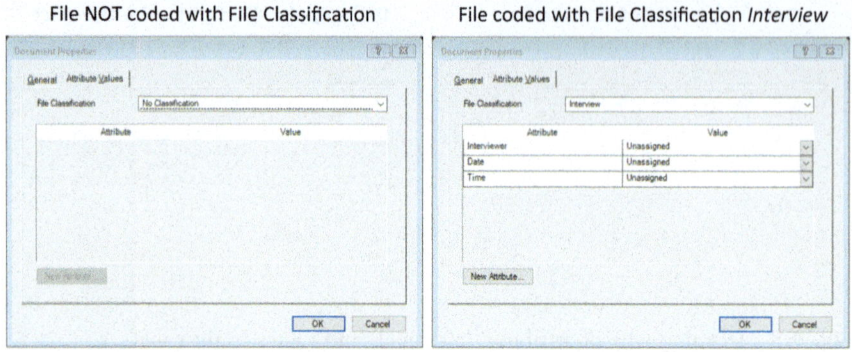

Fig. 9.9 Coding a file with a file classification

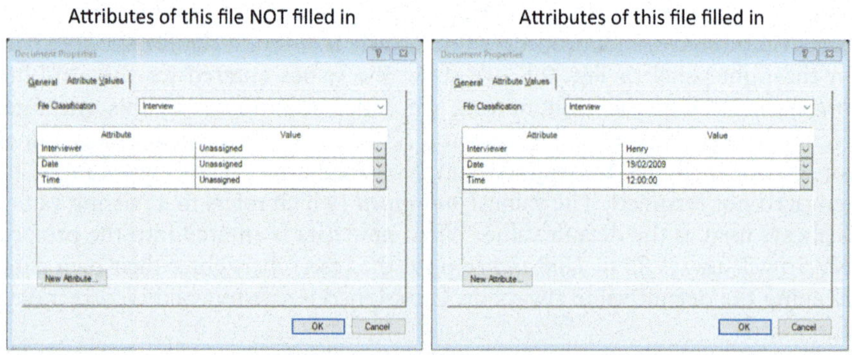

Fig. 9.10 Adding information to a file classification in a file

The moment you choose a file classification from the drop-down next to the option *File Classification*, the different attributes that are part of that classification appear underneath in the window. NVivo does not know the actual value of these attributes in this particular file and, therefore, selects the default value *Unassigned* to all attributes. STEP 3 now consist of filling in the data that belongs to this specific file. As shown in Fig. 9.10, the interviewer's name is *Henry,* and the interview was done on 19/02/2009 at 12:00.

The example shows how one file was coded with the file classification *Interview* and how data belonging to that particular file was added. We now have one file coded with complete information. You now need to repeat this procedure for all files: coding in the file's properties and filling in the data for each file.

Note on Classifications and the Navigation View

Before we go to the explanation of the Case Classifications, we first want to point out a particularity in the *Navigation View* that might confuse you if you are unaware of it. As we have explained (see Table 9.3), you can find the

File and Case Classifications in the *Navigation View* under **Data** and **Cases**, respectively. The *List View*, however, adapts depending on where you click in the classification sections. In Fig. 9.11, we explain this double view. When you select the folder *File Classifications* (or Case Classifications), NVivo shows the technical definition of the classifications available in the project. Four file classifications were created in the sample project, and in Fig. 9.11 (top panel), we show the details of the classification *Interview*. In the List View, you can see how this classification has three attributes: *interviewers, date* and *time*. Underneath this folder in the *Navigation View*, you see the names of the four file classifications repeated. When you click in the *Navigation View* on the folder *Interview*, the *List View* changes and now shows the data coded with this file classification. You now see that this File classification has been used in the interviews with Helen, Ken, etc. When you open the file with the ± sign, you get a view of the actual values filled in for that file (see the info on the interview with Barbara in Fig. 9.11, lower panel).

This double view on the technical definition or the data coded to the classification is programmed identically for the Case Classifications folder in the *Navigation View* under **Cases**.

Making and Using Case Classifications

The use of case classifications is very similar to that of file classifications. Therefore, some repetition in this paragraph is unavoidable. The step-wise work plan goes as follows:

Prerequisites before you start

- Files with data are imported into your project.

Work plan

STEP 1. Create the Case Classification (with Attributes and Values)
STEP 2. Create a twin case for each unit of analysis
STEP 3. Code your cases with the Case Classifications
STEP 4. Add the file-specific data to each coded case
STEP 5. Code your File(s) with the cases.

Creating the Case Classification (STEP 1)

As NVivo does not know which case classification we want to use, we again start with creating the instrument with its attributes. Making a new Case

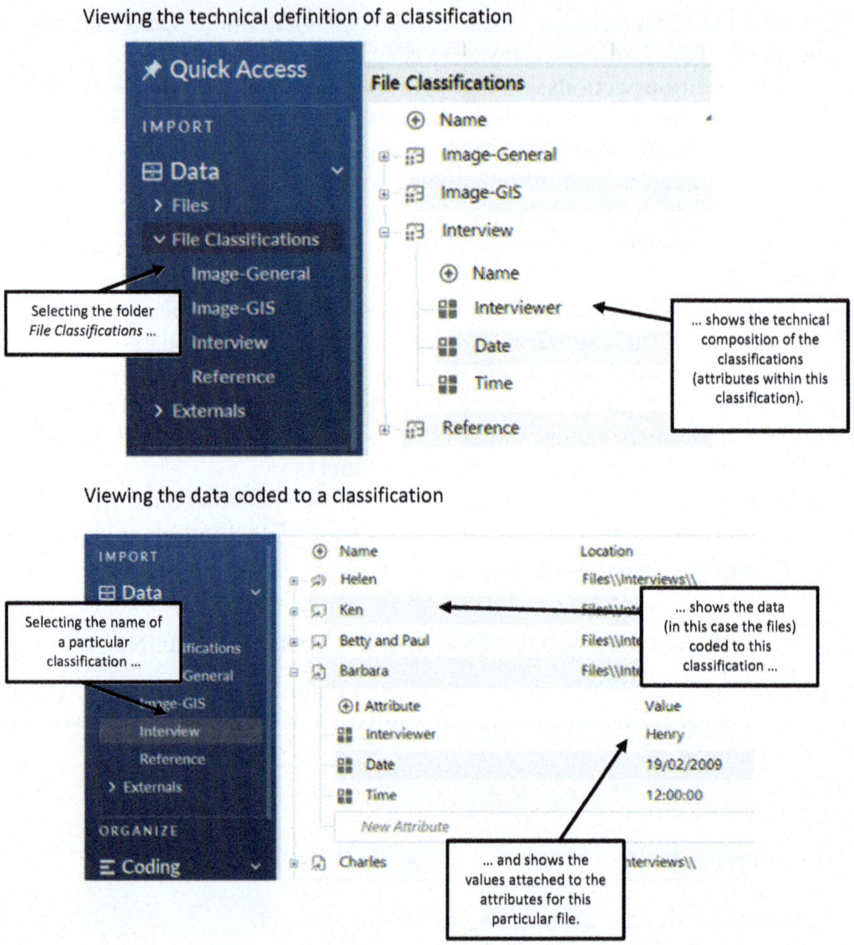

Fig. 9.11 Double view on classifications dependent on the selection in the *Navigation View*

Classification is done by **Create > Case Classification**. After giving the case classification a name, the instrument is created in **Cases > Case Classifications**. Attributes are then added by **RMC** (above the name of the Case Classification) > **New Attribute**. We refer again to Fig. 9.7, showing the classification and its attributes.

Creating Cases (STEP 2)

Case Classifications differ from File Classification mainly in the target to which they can be coded. As their names suggest, File Classifications are coded to Files and Case Classifications are coded to Cases. The concept of Files is clear:

Table 9.4 Icons of codes, case codes and cases in NVivo

Code (NVivo V14)	*Case-code (NVivo 12)*	*Case code (NVivo V14)*

these are data sources stored in the folder **Files** of our project. The idea of a *Case* needs some further explanation. The easiest way to explain the true nature of a case is to regard a case as a particular type of code. In Table 9.4, we compare the icons of previous versions of NVivo used for cases and the current ones. As shown in Chap. 8, the icon of a code in NVivo is a circle. The current icon of a case is a briefcase. But interestingly, the old icon of a case was also a circle filled with colours. This older icon showed that cases were a particular type of code. This idea is lost in the new icon while it is still very interesting for users to make this link: cases are *specific types of codes saved outside the codebook.* In the previous versions of NVivo, cases were also literally identified not as cases but as "case codes". This nuance has been lost in the current version, further disconnecting the current cases from the code idea. Nevertheless, it helps to see cases as a particular type of code that can bear the information of a case classification inside them.

Since we do need Cases to work with Case Classifications, we need to create these cases in our projects (contrary to the files in the previous paragraph that were already imported into our project before we started working with File Classifications). But which cases need to be created, and how many of these cases need to be created? The answer is simple: consider a case as a twin for your unit of analysis. So, for each unit of analysis, you create one case as a twin (to which we will refer as twin cases). And since you need to be able to code your material with these twin cases (they are special codes, remember), it is good to name them with the same name as the unit they refer to. So, if you have an interview with a respondent called Barbara, you create a twin case for Barbara to code her material. This idea is shown in Fig. 9.12: for each interview participant, there is one twin case with the same name.

To create a case, go to **Create > Case**. Cases are project items, so they need a name. As explained, choose the name of the unit of analysis this case refers to. This twin-character of cases is essential for the coding in STEP 5.

Code the Case with the Classification and Add the Case-Specific Data (STEPs 3 + 4)

When you create a new case, you are asked to give the case a name in the tab General. When you go to the tab Attribute Value, you can code the case with

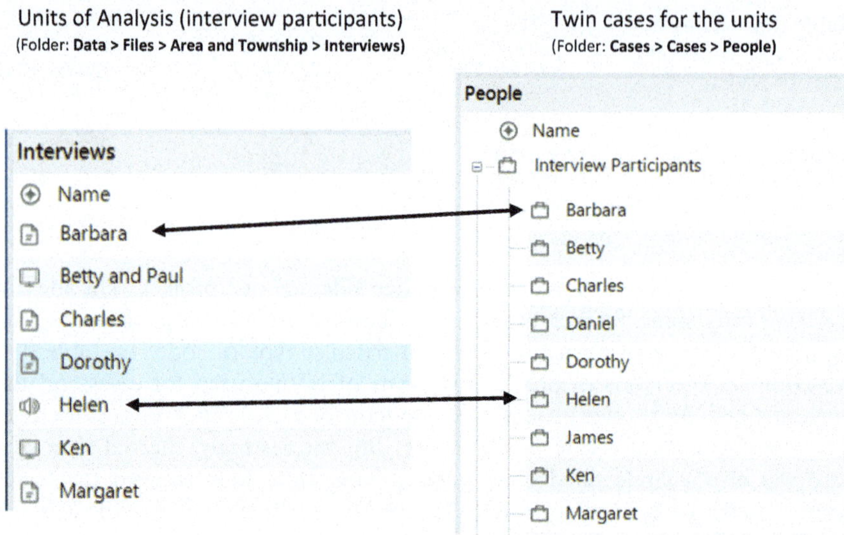

Fig. 9.12 Comparison of units of analysis and twin-cases for each unit

the Case classification. When you have closed the case, you can re-open the case with **RMC** (above the name of the case) > **Case Properties**.

Coding the case with the Case Classification is similar to the File Classification: you use the drop-down next to the option *Case Classification* to select the classification instrument you want to use with this case (see Fig. 9.13). Next, you fill in all the values of the attributes with the information that belongs to this unit of analysis (see Fig. 9.14).

You repeat this process of attaching the classification to the cases and filling in the information of the cases for all twin cases you have made.

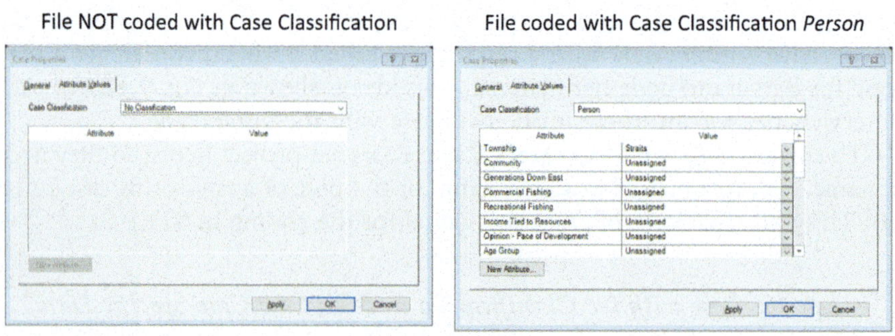

Fig. 9.13 Coding a case with a case classification

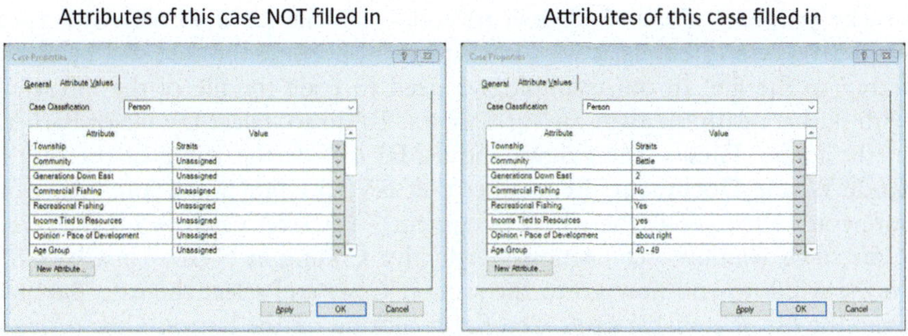

Fig. 9.14 Adding information to a case classification in a case

Code Your Data with Their Twin-Case (STEP 5)

The work was done after coding a file with a file classification (see STEP 3 in Section "Using the File Classification (STEPs 2 and 3)") because these files contain your data. So, if the classification information is saved inside your file, your data is immediately coded with the classification information. This is not the case with the case classification. If the case is coded with the case classification and the information is filled in, you have a case with information in your project that is *not* attached to your data. In Fig. 9.15, you see this in the list of cases and files when you coded your case with the case classifications.[1] Neither in the list of cases nor the list of files is there any proof of a connection: 0 Files and 0 References in both lists.

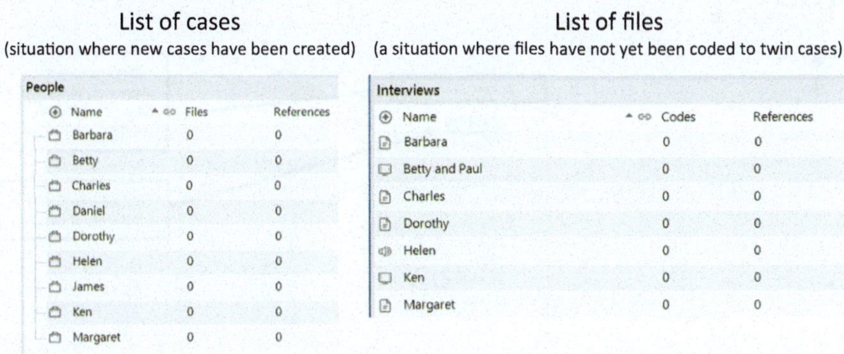

Fig. 9.15 Cases not attached to files (left panel) and files not coded with cases (right panel)

[1] In this example, we have recreated the list of cases and files to show a situation where a user has created new cases and has not yet coded the files with the cases. In the sample project, this is all done so the screenshots are not taken directly from the sample project. The example also simulates the fact that there was no thematic coding done yet in this project.

The purpose of step 5 is now to code each file (as a whole) with their respective twin case.[2] That way, the information inside the case will be transferred to the file. In our example, we need to code the file of the interview with Barbara with the twin case of Barbara. To do so, select the file of Barbara in the folder **Files > Interviews** and **RMC** (above the name of the file) > **Code Whole Document**. Remember that NVivo might warn you, "Are you sure you want to code the whole document?" Click on *Confirm* for the *Select Code Items* window (see Section "Deductive Coding in NVivo" in Chap. 8). In the window, you now go to the section *Cases* and select the case *Barbara* from the list of available cases. You finish clicking on the button *Code selection to 'Barbara'* (see) (Fig. 9.16).

The result of this coding is shown in Fig. 9.17: in the list of Cases (left panel), NVivo shows that the case Barbara is now coded to 1 File and 1 Reference (the whole file is the reference in this case). In the list of Files (right panel), the interview with Barbara is now coded to 1 Code, again with the same 1 Reference. In this list, it again turns out that NVivo considers cases as codes since the coded case is adopted in the column count of *Codes*.

To close this paragraph, we show in Fig. 9.18 a screenshot of the opened interview with Barbara. When you activate both the Coding Stripes and the Code Highlighting, you see that the case 'Barbara' is also considered a code, represented by one long coding stripe across the whole interview text. Also, the text of the interview is entirely coloured yellow since the entire document was coded with this case. Graphically, there is no difference between

Fig. 9.16 Coding files to cases

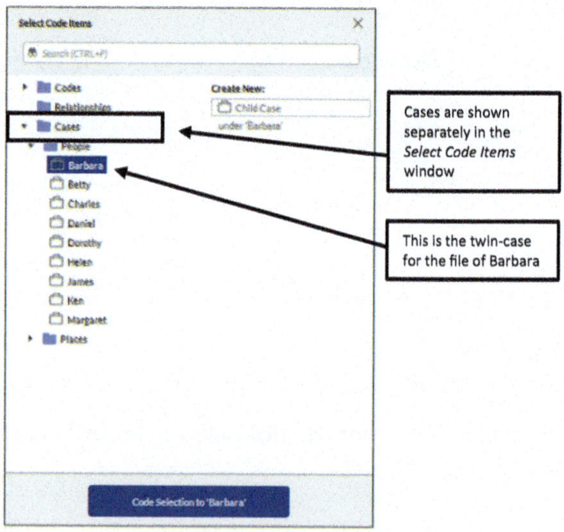

[2] Caveat: This is the situation where you work with interview data. For focus groups, this is not necessary the case as the levels of analysis may differ there. See Section "Using Classifications in Focus Groups" in Chap. 15 for more information.

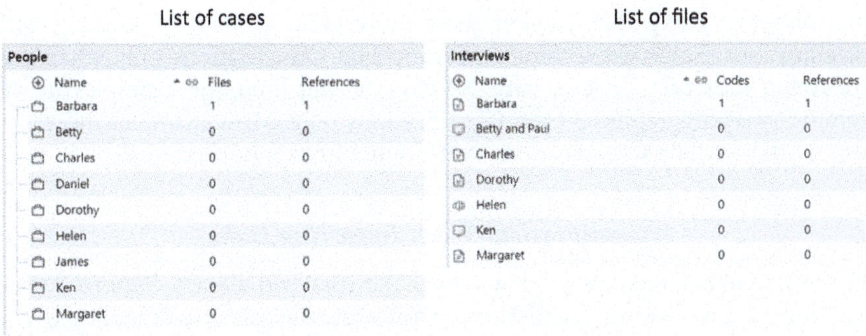

Fig. 9.17 File 'Barbara' got coded with the case 'Barbara'

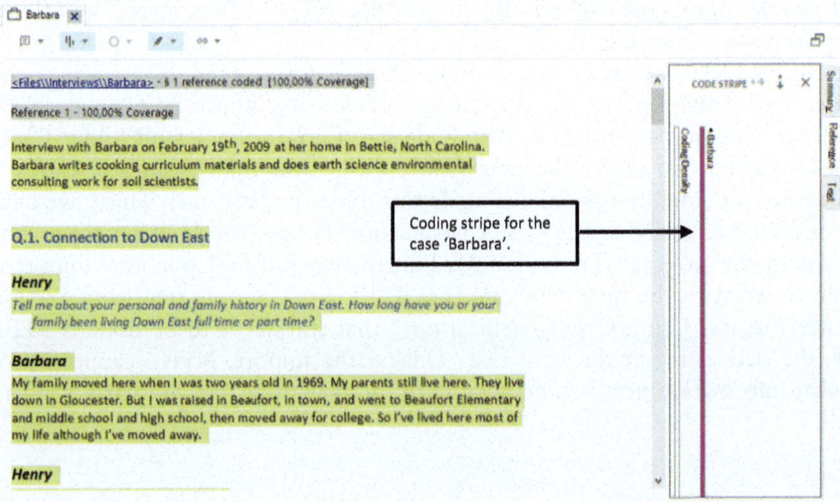

Fig. 9.18 File 'Barbara' coded to the case 'Barbara' (opened interview view)

coding stripes of cases and coding stripes of thematic codes. So, when you add thematic coding in the interview with Barbara, the differences between codes and cases will not be apparent from the *Coding Stripes* panel.

Gaining Efficiency in Creating and Using Classifications

In Sections "Making and Using File Classifications" and "Making and Using Case Classifications", we presented the default way of working with File and Case Classifications. We also aimed to show you the fully manual way step by step (e.g. not using pre-programmed classifications). NVivo has many shortcuts that make the work with these classifications more efficient. This and the following paragraphs present these more efficient ways of working with classifications. Since we recommend using File Classifications only when you are doing literature reviews with NVivo, we will limit the examples in the coming two paragraphs to Case Classifications only. Most examples, however, also work with File Classifications (with only minor modifications).

Classifications When Importing Data

In the previous paragraphs, we assumed the user had already imported their data files. To work with classifications, however, this is not a necessity. If you use Case Classifications, it might be more efficient to add the classification when you import files or data immediately.

When you import data (**Import > Files**), you can tick the option *Create a case for each imported file*. When you do, NVivo offers several additional options (shown in Fig. 9.19).

In Fig. 9.19, we ask NVivo to do much work for us. First, NVivo will create a case with the same name as the name of our new imported file (i.e. STEP 2). Second, we have this new twin code saved not in the default folder **Cases > Cases,** but we divert the new twin code to our folder **Cases > Cases > People** (provided that you have made this folder beforehand). Third, we have our twin case coded to the Case Classification *Person* (which also must already exist in the project) (i.e. STEP 3). Fourth, we will find our new imported file coded with the new twin case (i.e. STEP 5). The only step from Section "Making and Using Case Classifications" that still needs to be done is STEP 4, the data entry in the twin case. During the import, NVivo cannot know what information needs to be filled in in the new twin case. So, what is left to

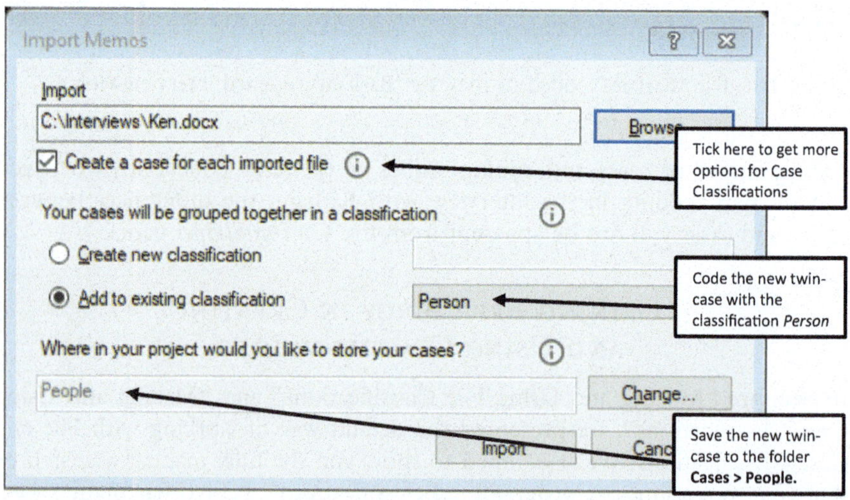

Fig. 9.19 Using case classifications while importing files

do for the user is to open the properties of the twin case (**RMC** (*on the name of the twin case*) > **Case Properties**) and fill in the data of this unit of analysis.

Cases from Files

You can save time by automatically creating twin cases when you have imported your files into your project without using the procedure in the previous paragraph. When you select all files in a folder that need to have twin cases, you can **RMC** (above the selection of file(s)) > **Create as** > **Create as Cases** (i.e. STEP 2). NVivo shows the window *Select Location*. In this window, you can select the folder where the new cases need to be created and indicate the Case Classification to which these new twin cases need to be coded (i.e. STEP 3). NVivo will automatically code the file(s) with the new twin case(s) that were created (i.e. STEP 5). Again, the only thing left for the researcher is to open the properties of the twin case(s) and fill in the data that belong to these respective units of analysis (i.e. STEP 4).

Using the Classification Sheet for Data Entry

In the previous two paragraphs, we showed efficient ways to create twin cases and the attachment of these twin cases to a case classification and the data. In both cases, the last step (i.e. STEP 4) was not performed automatically. We referred to the manual data entry in these cases by **RMC** (on the name of the twin case) > **Case Properties** to fill in this data in the twin cases.

A more efficient data entry method is available with the Case Classification sheet. The Case Classification sheet is a tabular representation of all cases (in the rows) and all attributes (in the columns). It is opened in a separate tab by **Home** > **Case Classification Sheet** > **(name of the case classification to be represented in the sheet)** (see Fig. 9.20).

The cells of this table are editable and can be used for data entry. When you need to enter data in a large number of cases, entering the data in the

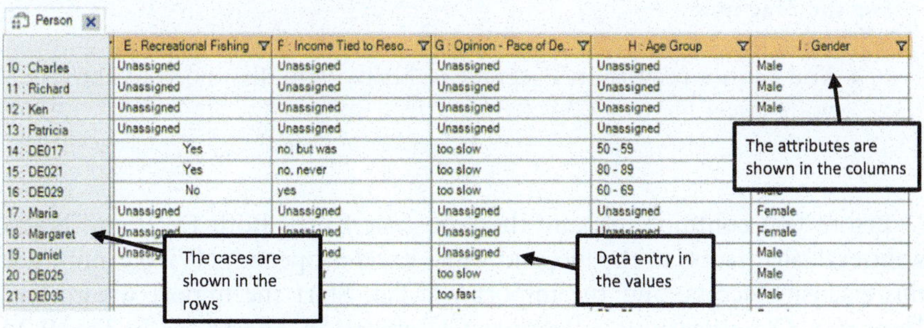

Fig. 9.20 Case classification sheet

case classification sheet might be more efficient as you get a general overview of the cases and the attributes.

The classification sheet can also be exported to an Excel sheet of an SPSS dataset by **RMC** (above the classification sheet) > **Export Classification Sheet**. This allows further (quantitative) analyses of the data in your project.

Fully Automated Use of Classifications: The Descriptive Matrix

The last possible gain in efficiency when working with classifications starts outside NVivo with the so-called *descriptive matrix*. Many researchers keep track of their respondents in an administrative database. They register names, addresses and background information. Or they ask their respondents to fill in a drop-off, i.e. a short survey with closed questions on their background. They keep track of all this information in Word, but the data is often stored in an Excel sheet. NVivo allows the researcher to gain efficiency by importing this Excel sheet into their project and creating classifications with it. By doing so, all five steps we have explained earlier (see Section "Making and Using Case Classifications") are all done automatically.

Before we explain the import, let us concentrate on the Excel sheet. This must be in a very specific format for the import to work. In Fig. 9.21, we show how the different components of the sheet need to be prepared. In cell A1, you put the name of the Case Classification that NVivo will create. On the first row (from column B onwards), you put the names of the attributes that need to be created within the new case classification. In the first column (from row 2), you put the names of the cases that need to be created. If these names are *identical* to the names of files (i.e. twin cases), NVivo will also code these files with the twin cases. In the other cells, the data for each case on each attribute is recorded. Each case is a row so that you can read the data of one particular case along the rows.

For the Mac user

- You cannot import an Excel file in NVivo. You first need to convert your Excel file to a CSV file.

Before you continue, be sure that you *close* Excel. Both Excel and NVivo want exclusive access to the file, and while Excel is opened, NVivo cannot gain this exclusive access, and the import will fail. Start the import wizard with **Import > Classifications > Import Classification Sheets ...**. In Fig. 9.22, we give all four import wizard steps and show you which options you need to select to import the Excel file.

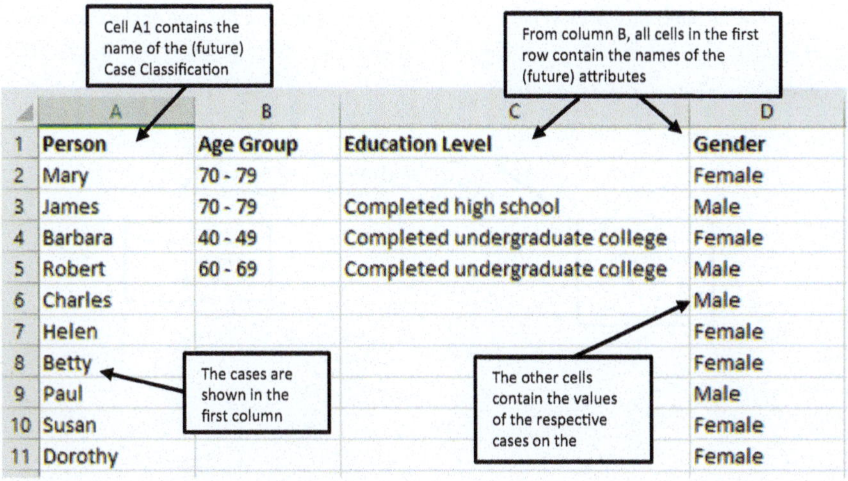

Fig. 9.21 Descriptive matrix in Excel

When you click *finish* in step 4, NVivo will import the descriptive matrix. All five steps from Section "Making and Using Case Classifications" are now done automatically: a new Case Classification (with Attributes and Values) was created in your project (STEP 1). NVivo took the content of cell A1 as the name of the new Case Classification. For all data in the first column, new cases are created at your specified location (STEP 2). All the new cases are coded with the new Case Classification (STEP 3), and all values from the table are automatically filled in for each case (STEP 4). When the names of your cases match *identically* with the names of the files, NVivo has also coded these files with their twin cases (STEP 5). This last step, however, sometimes fails as NVivo is very strict in the link between twin cases and files. When a minor detail seems different, the file is not coded.

When you look closely at the options in Fig. 9.22, you can see that several options concern the update of the classification and the cases. The wizard is constructed as a tool that researchers can use regularly. So when you do interview research, you can start importing the descriptive matrix when you only have two interviews and later update the classifications, the cases and the coding when you add rows in your Excel file. Just import the same Excel file again, and the elements in your NVivo project get updated every time. That makes the import wizard for classifications a very powerful tool in your work.

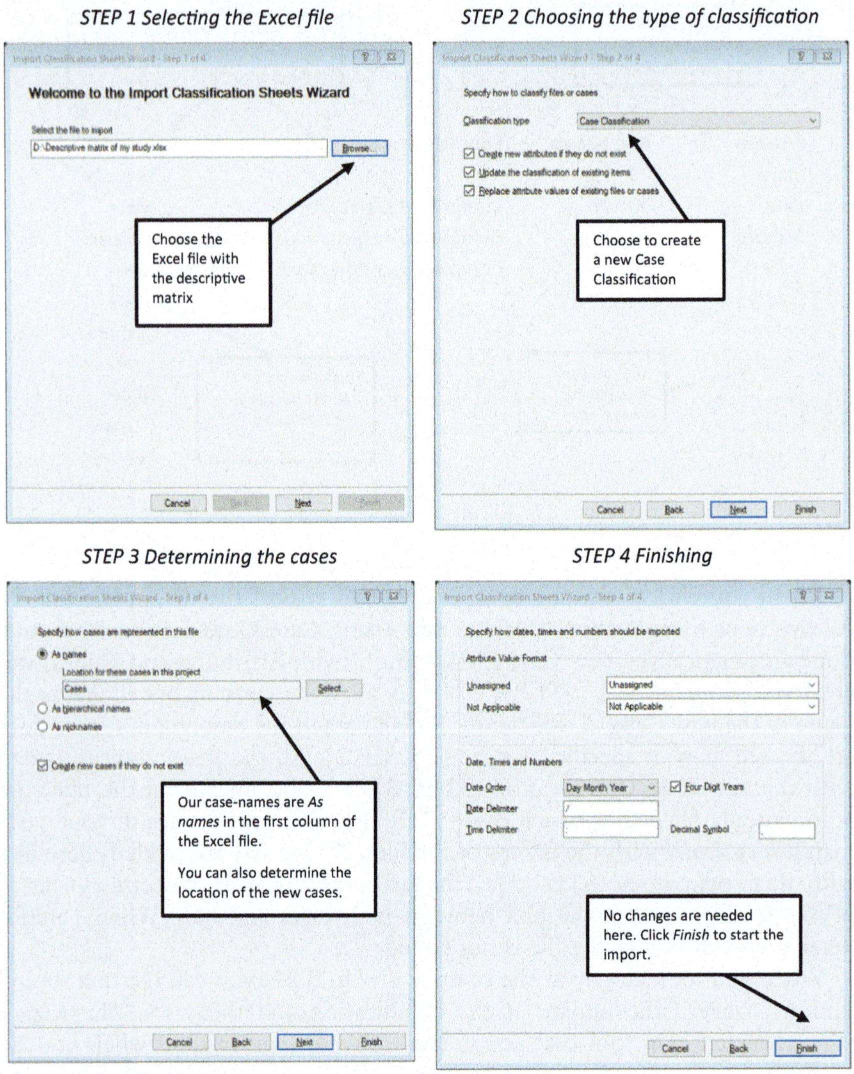

Fig. 9.22 Importing a case classification sheet from Excel

CHAPTER 10

Exploring Coded Data

Key messages in this chapter

- The Explore diagram visualises the way project items are connected.
- The Comparison Diagram allows you to compare two project items in their relatedness.
- The Framework matrix helps the researcher make summaries of their coded data.

In this chapter, we present three unrelated tools that all assist you with interpreting your results after your data is coded (or from the moment a substantial part of your data is coded). Some tools will fit the analysis style of some researchers better than others. Therefore, look at the tools, try them and see what works for you and what does not. None of these three are indispensable in analysing your qualitative data, but they all have their individual (visual) power.

Walking Through Your Data with the Explore Diagram

The Explore Diagram is a dynamic diagram that allows you to virtually "walk around" through your data. The diagram shows a project item (e.g. an interview, a code, a case) in the middle of the diagram and visually shows all the other project items this item is connected to. You can create new diagrams

D. Mortelmans, *Doing Qualitative Data Analysis with NVivo*, Springer Texts in Social Sciences, https://doi.org/10.1007/978-3-031-66014-6_10

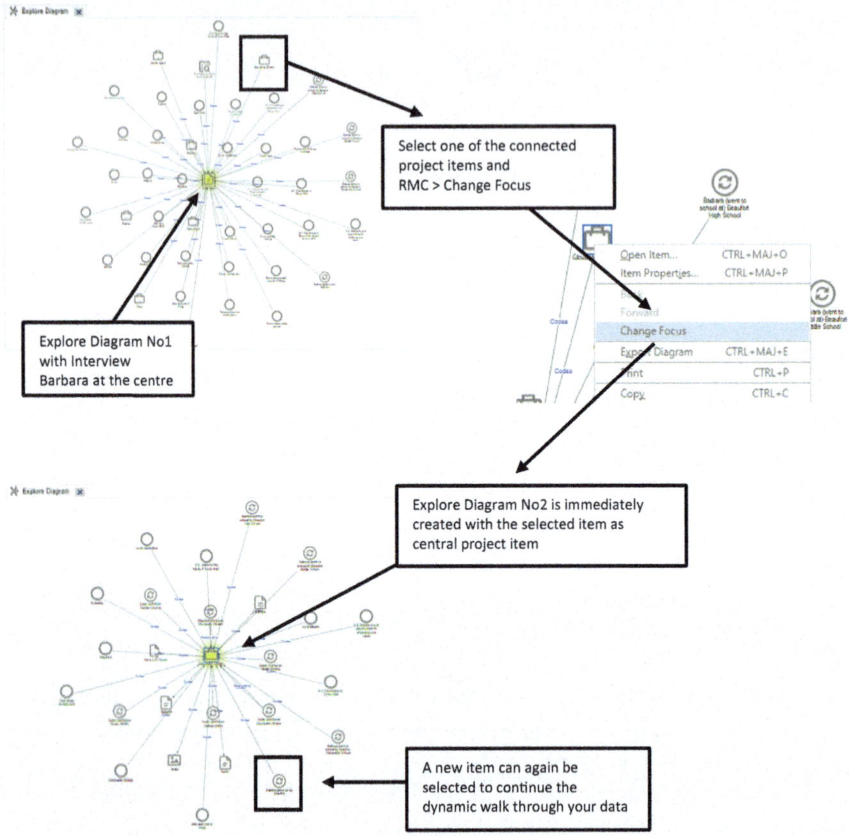

Fig. 10.1 Dynamic use of the explore diagram

from this first diagram by selecting one of the attached project items and placing that item in the centre of a new diagram that replaces the first one. From there, one can again create a new diagram from one of the newly shown attached items. And so on.

We illustrate this dynamic process in Figure 10.1. To create the first diagram in the figure, select the interview with Barbara in **Navigation View > Data > Files > Interviews**. Next, go to **Explore > Diagrams > Explore Diagram**. You now see the Explore Diagram with the interview with Barbara at the centre.

The Explore Diagram can be exported by **RMC** (above the diagram) > **Export Diagram**. Export possibilities to all major graphical formats (e.g. jpg or png) are available. The diagram can also be printed similarly. When the diagram contains too many elements to be interpretable, you can use the checkboxes in the **Explore Diagram** menu to show or hide certain project items from the diagram.

Dual Comparisons with the Comparison Diagram

The Explore diagram shows you one project item and its connectedness to other items in your project. The Comparison Diagram allows you to compare the connectedness of two project items. Three variants are available: a comparison of files, a comparison of codes and a comparison of cases. You will find these three possible diagrams under **Explore > Diagrams**. A fourth possibility is available when you select two of the mentioned project item types (e.g. you select two files), then NVivo shows the option **Compare** in **Explore > Diagrams**. When you select only one project item (e.g. one case), the previous options are changed in **Explore > Diagrams > Compare With,** and you can select a second project item (see further) to compare your selected item with.

In Figure 10.2, we show a Comparison Diagram comparing two sample project interviews. We started creating this Diagram by **Explore > Diagrams > Compare Files**. In the window *Select Project Items*, we selected the interviews of Barbara and Charles in the folder *Files > Interviews*.

Remark

- In the window *Select Project Items*, the button OK only becomes selectable when you select two and exact two project items to be

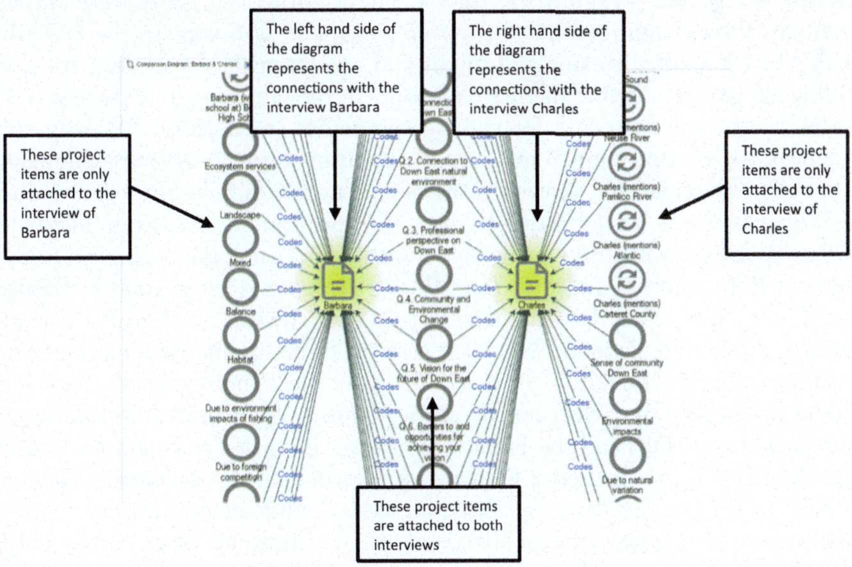

Fig. 10.2. Comparison diagram (zoomed-in representation)

compared. Selecting fewer or more items will not enable you to make the diagram.

Analogue to the Explore Diagram: the Comparison Diagram can be printed or exported by right-clicking above the diagram.

Making Summaries with the Framework Matrix

In Chap. 12, we will present how a researcher can retrieve the coded references from their data. The researcher will then read and interpret these references and analyse them further. However, many researchers desire to summarise their data in a structured way to get an overview of larger themes arising in their analysis. The Framework Matrix is a tool that allows you to make (and save) summaries of coded references in your project.

For the Mac user

- The Framework Matrix is not yet available in the MAC version.

Unlike the Explore and Comparison Diagrams, the Framework Matrix is a stand-alone project item. This means we must create a new Framework Matrix in **Create > Framework Matrix**. The window *New Framework Matrix* contains three definitional tabs: *General, Rows,* and *Columns*. In the first tab, you give the matrix a name and might add a description if you want to. The following two tabs define the matrix layout. In the tab *Rows*, you place several cases that you want to use for making summaries (see Section "Making and Using Case Classifications" in Chap. 9 for using cases in NVivo). Strangely enough, only cases are allowed as rows and not Files (e.g. interviews). But as you can easily assign cases to files (see Section "Cases from Files" in Chap. 9), this should be a manageable threshold for using the tool. Next to the section *Rows*, there is a section *Row Header Attributes.* Here, you can choose attributes from a Case Classification to sort your rows in the final Framework Matrix. In the tab *Columns*, you place content to be summarised, i.e. thematic codes from your codebook. In both tabs, you select the respective cases and codes by clicking the Select button at the bottom. When selections have been made, you click OK, and the Framework matrix is created as an empty matrix. In Figure 10.3, we created a Framework Matrix named "*Summaries Sense of Community*". In the rows, we place all cases found under *Cases > People > Interview Participants* and in the columns, we chose all three codes under *Codes > Sense of Community Down East*.

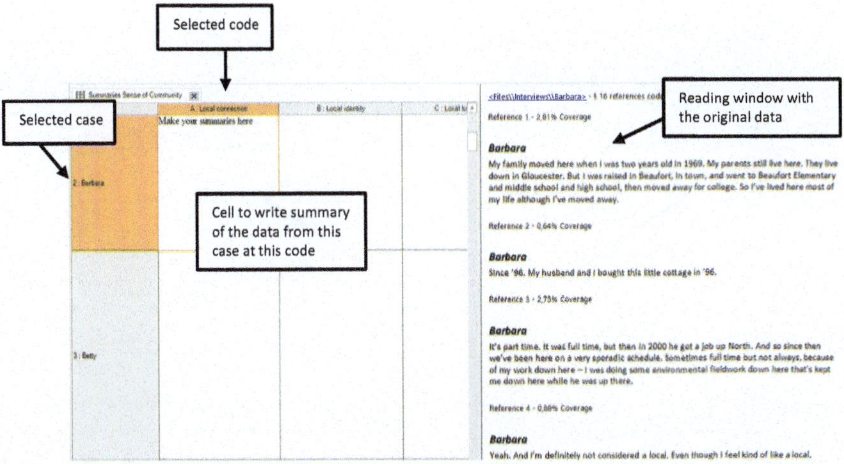

Fig. 10.3. Framework matrix of interview cases (rows) and thematic codes "Sense of Community Down East"

All cells are empty when the Framework Matrix is created (at the left side of the window). On the right-hand side, the coded references are shown so the researcher can start summarising the data in the empty cells. In the reading window, all references are shown for the selected case (row) and not only for the selected code (column). To change this, you can limit the content of the reading window to the references that relate to the selected cell by **Framework Matrix > Associated View > Cell Coding**. This makes the interpretation (and summary) easier as now you only see the original data that is coded to the selected cell only. If you click on a cell and the reading window is empty, you know that this case has no coded information on that particular thematic code and that a summary is impossible.

Remark

- The Framework Matrix is not the same as the Matrix Coding Query. The Framework Matrix allows the making of summaries of themes, while the Matrix Coding Query bundles references according to the search criteria of the query (see Section "The Matrix Coding Query" in Chap. 12).

One step further is the opportunity to copy all original data from the files to the respective cells of the Framework Matrix. As such, all cells where references are found are placed in these cells, and the researcher can transform the original data immediately into summaries. This is a bit misleading, you can

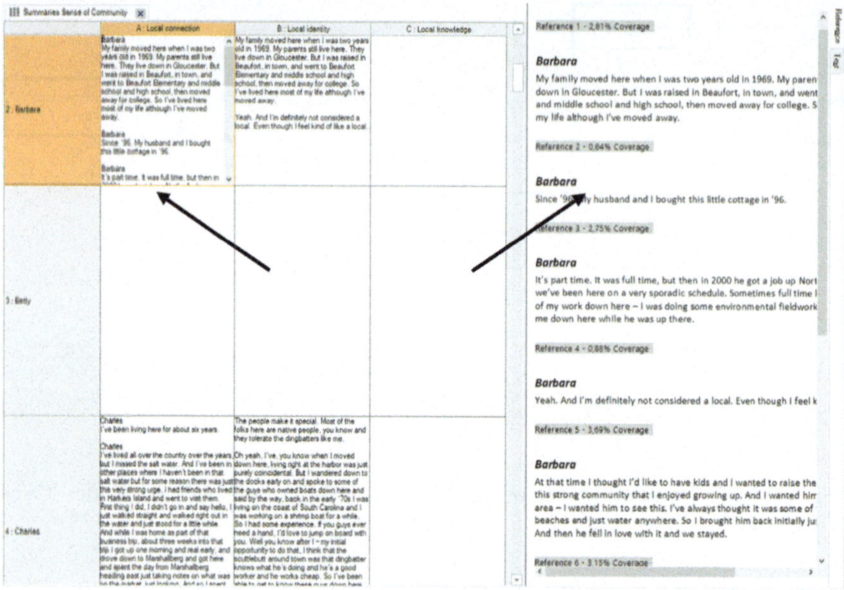

Fig. 10.4 Framework matrix after applying the *Auto Summarize* function

find this feature under **Framework Matrix > Auto Summarize**. Do not be deceived by the name of the function: NVivo does *not* make the summaries for you. It only copies the original data to the cells of your Framework Matrix, and then *you* need to make the actual summaries. In Figure 10.4, we illustrate the result of this function to the Framework Matrix we used as an example in Figure 10.3.

To conclude, the Framework matrix can be printed or exported by right-clicking on the matrix. Export possibilities contain both the Excel format and text file format.

CHAPTER 11

Organising Your Project: Secondary Data Management

Key messages in this chapter

- Relationships allow you to link project items in your project.
- See-also-links allow you to make referrals in your data.
- Sets (static and dynamic) group items in your project so that you can select project items.

In the first chapter on data management (Chap. 6), we introduced the basic setup of an NVivo project: setting up a folder structure, importing data and handling your data files. In this chapter, we introduce data management techniques when the project is in a more advanced stage. When most of your data files are imported and coded, you can link project items with relationships and see-also-links. In addition to grouping project items with folders (see Section "Setting Up a Folder Structure" in Chap. 6), different groupings of project items can be realised with static or dynamic sets. Unlike the primary data management chapter, importing files, for example, is crucial to start your analysis. The tools in this chapter are more specialised. Only some types of analysis will require these more specialised tools, but the researchers who do will undoubtedly appreciate their availability.

D. Mortelmans, *Doing Qualitative Data Analysis with NVivo*, Springer Texts in Social Sciences, https://doi.org/10.1007/978-3-031-66014-6_11

Linking Data with Relationships

Relationships in NVivo are considered to be a specific type of code. Therefore, we could have discussed the tool in Chap. 8. But the resemblance with see-also-links (see the next paragraph) made us decide to introduce the tool in this chapter.

Relationships make connections between project items. They allow the researcher to identify a relationship between items. Typically, relationships are used to link two codes to show the relationship between concepts, or they link two files, e.g. to show the connection between participants in the data.

Since NVivo does not know what relationships the researcher wants to use, you must first create "relationship types" in your project. This implies that you create a stock of relationships in your project to be available for linking your project items. In the second step, you will link two project items with your library of relationship types. Both components are visible in the *Navigation View* > *Coding* (see Fig. 11.1).

Step 1. Creating Relationship Types

First, you must prepare to use relationships by building a library of relationship types. Comparable to creating a (case or file) classification where you start with making attributes, you here need to start with creating the relationship type (by giving it a descriptive label) before you can actually use this relationship in your project. NVivo allows you to define three types of relationships:

One way	→	Barbara *went to school* Beaufort Elementary School
Associative	—	Mary *is married to* James
Symmetrical	↔	Community change *contributes to* Environmental change

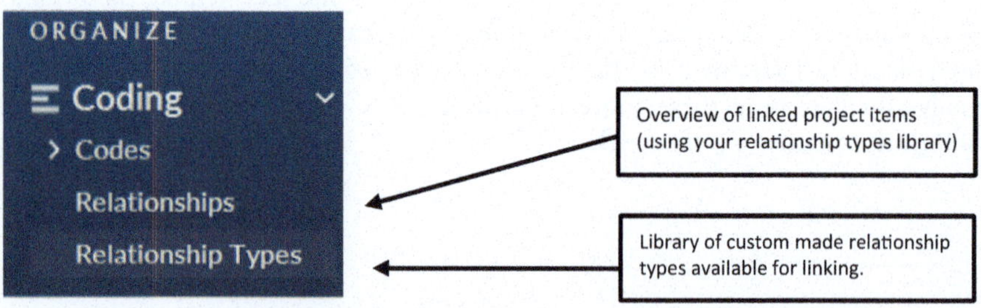

Fig. 11.1 Sections "relationships" and "relationship types" in the navigation view

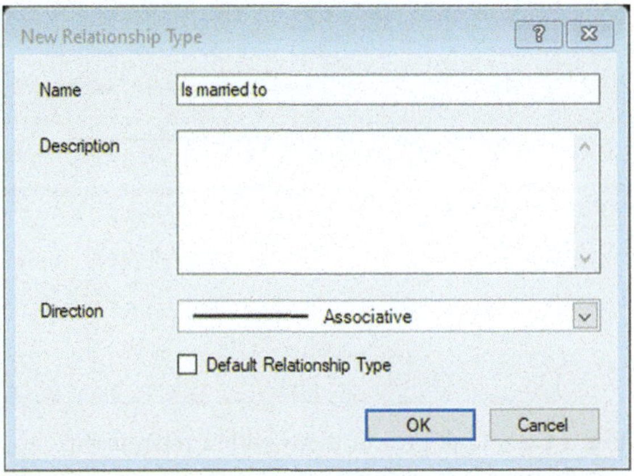

Fig. 11.2 Creating a new relationship type

When defining a relationship type like "went to school", the researcher needs to determine the nature of the relationship: one-way, associative or symmetrical. As "went to school" is a clear one-directional relationship, this relationship type was defined as a "one-way" relationship visually identified with the one-side arrow.

In order to create a new entry in your relationship library, you select *Navigation View* > ***Coding*** > ***Relationship Types*** and you **RMC** (*in the detail view window*) > **New Relationship Type ....** Defining the relationship type is relatively straightforward (see Fig. 11.2). In the field *Name,* you label the relationship. In a concise manner, you define what the relationship entails. You could give more detail in the field *Description* (optional) and choose one of the three types above in the field *Direction.* NVivo also allows you to turn this relationship type into the default type. When creating the project, you will see that there is always one relationship type present in your project (named "Associated"). This relationship, "Associated", is the default type in a new project, but the researcher can determine the default type while creating new types.

When clicking OK, NVivo creates the new relationship and lists it in the *List View*. The process is repeated for each new relationship type the researcher wants to use in their analysis.

Step 2. Assigning Relationships to Project Items

After creating relationship types, the relationships can be used to link project items during the analysis. To link two project items, go to **Create > Relationship**. As shown in Fig. 11.3, you choose the project items with the *Select* buttons next to *From* and *To.* In the example, we have selected the cases of Mary and James, both located in Cases\People\Interview Participants. The drop-down in the middle shows a list of all relationship types available in your relationship library. From the list, we selected *Is married to,* to indicate that the

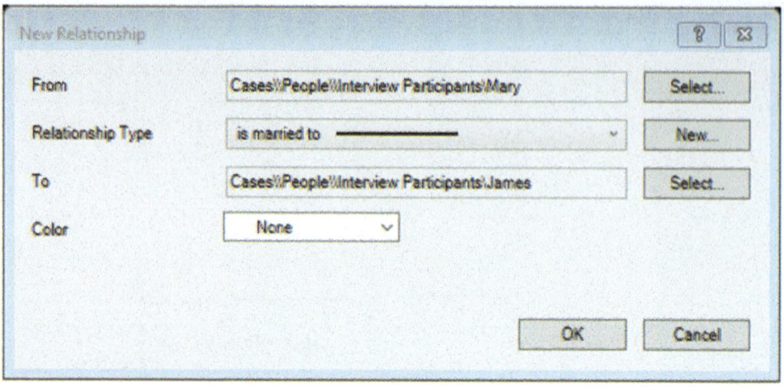

Fig. 11.3 Link project items with a relationship

participants, Mary and James, in our study are spouses. If you miss a specific relationship in the list that you want to use, you can immediately create a new type from this window using the *New* button. NVivo will then show the window from Fig. 11.2.

In the example, we link two cases in the sample project. This is only one possibility of many. You can also link two Files, two codes, two memos, or a combination of any of these. The linkage possibilities are very flexible in NVivo, and the researcher has sufficient freedom to link whatever content they see matching in the project. *Navigation View* > **Coding** > **Relationships** provides an overview of all links made with relationships.

Linking Data with See-Also-Links

See-also-links are also a tool to link information in your project but on a different level than relationships. Relationships link project items and serve as connections between these items. See-also-links connect content *within* project items and function on the intra-item level. Another difference with relationships is that see-also-links just link two pieces of content without defining what the link actually is. When using relationships, the researcher has to specify the type of relationship between the project items (e.g. *is married to*). For see-also-links, such a "type of link" is not foreseen. You link two pieces of content, allowing you to jump between these two places in your project quickly. In this way, you can consider relationships as coding because the relationship is labelled. The see-also-link is more of a memo-like tool that connects information in your project items without labelling the content of that link.

There are two ways of linking content with see-also-links. The first method is linking content to an entire project item. The example in Fig. 11.5 links a fragment in the memo "Local identity and knowledge" to the File with the interview transcript of Barbara. The first step in creating this link is selecting the content in the memo. Next, you find an icon for See-Also-Links in the

Small ribbon of the memo document: ▾. In the drop-down, you can choose *Add See-Also-Link.* In the window that appears, the From-field has now been identified. With the Select button, you need to identify the To-field. You can now link to Files, Codes, Cases, Memos, etcetera. NVivo allows you to select one project item, shown in the *Item* field under *To.* In the example in Fig. 11.4, we selected the interview transcript of Barbara under *Files \ Interviews.* In the *List View* of *Navigation View* > **Notes > See-Also-Links,** a new entry is added showing both the From-item and the To-item. When you **RMC** (*above the See-Also-Link entry*), you can open the *From*-item or the *To*-item.

The third screenshot in Fig. 11.4 clearly shows that the option Entire Content is greyed out in the To-section of the window. With the method described before, making a See-Also-Link from content to content is impossible. A slightly different method is required to link two elements within a File. The second method starts again in the From-item by selecting the content that needs to be linked (see the first screenshot in Fig. 11.5). Instead of using the See-Also-Link-icon, you now open the To-file in your *List View.* In that file, you select the content to which the first content needs to be linked. The final step now consists of **RMC** (*above the selected reference*) > **Paste as See-Also**

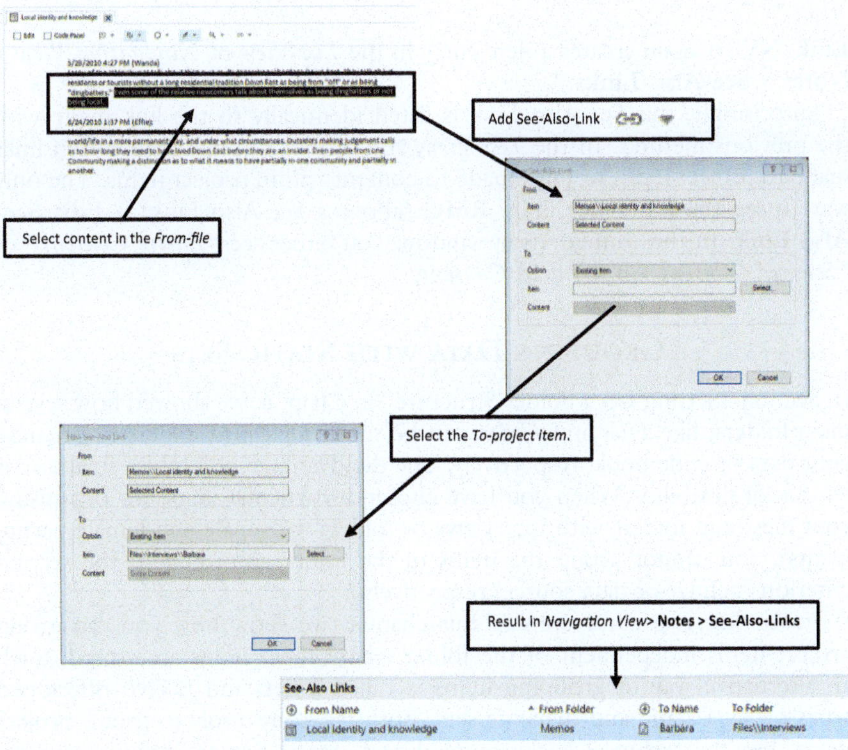

Fig. 11.4 Linking content with see-also-links to project items

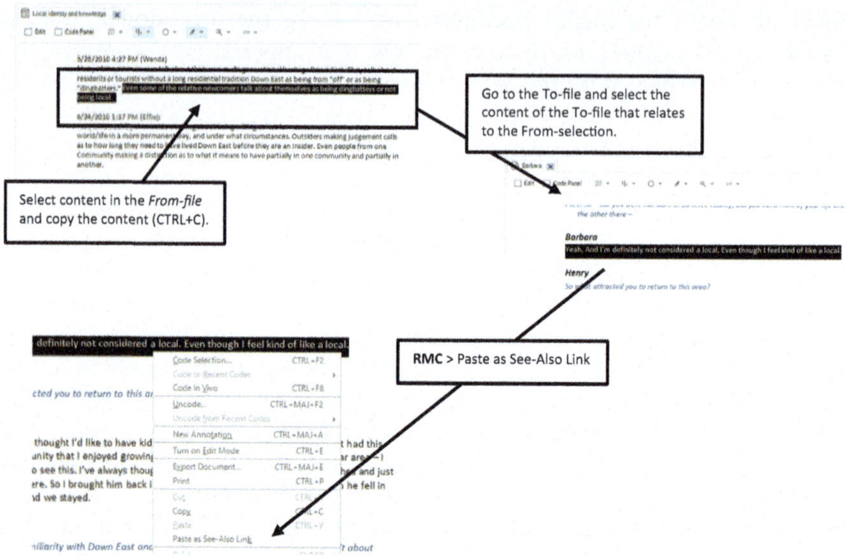

Fig. 11.5 Linking content with see-also-links to other content

Link. NVivo again creates a new entry in the *List View* of *Navigation View* > **Notes > See-Also-Links**.

Surprisingly, this See-Also Link is listed identically to the link made with the previous method. In the *List View*, you cannot distinguish between links made to project items or links made to content within project items. The only way to see the difference is by **RMC** (above a See-Also Link) > **Edit See-Also Link**. In the configuration window, you either see "*Entire Content*" or "*Selected Content*" in the field *Content*.

Grouping Data with Static Sets

In Section "Setting Up a Folder Structure" in Chap. 6, we showed how several main folders, like *Files* and *Codes*, can be subdivided into folders to organise your data or code book, respectively. The disadvantage of a folder structure is its lack of flexibility. When you have chosen a particular angle for organising your files (e.g. folders with interviews or folders with male and female participants), you cannot group the items in the folder according to the second criterion (gender or data source, respectively).

In the last two paragraphs of this chapter, we show how you can group project items independent of the folder where these items are stored. Such an alternative way of grouping items is called a Set, and NVivo offers two types of Sets: static and dynamic sets. Both types are tools to group project items, but the technical background differs. *Static Sets* are groups of items where the user manually adds project items to a group. These items stay in the group as long as they exist in the project. When new (and similar items) are added to the project, they are not automatically added to a static set. The researcher always needs to add new items manually. On the other hand, a *Dynamic Set* is a search operation for project items of which the results are stored in a group. This set is dynamic in that the researchers define a search, and the program determines which items are retrieved with the search and which project items are grouped in the set. When new items are added to the project that also complies with the criteria of the underlying search, these items are automatically added to the dynamic set. A dynamic set is constantly and automatically updated because every time the researcher opens the set, its members are searched for in the project at that instance according to the underlying search that was initially defined. In this paragraph, we explain static sets and refer to Section "Grouping Data with Dynamic Sets" in Chap. 11 for dynamic sets.

Create > **Static Set** > **New Static Set** creates a new static set. You give the Set a name, which is saved as a new entry in *Navigation View* > **Sets > Static Sets**. You add project items to this Set by **RMC** (*on the name of the Set* in *Navigation View* > *Sets* > *Static Sets*) > **Add Static Set members**. NVivo then shows the *Select Project Items* window, and you can select all project items that need to be grouped in this set.

A second way of creating a static set is by selecting some project items (e.g. you select a few files or a few codes). You can now immediately create a new Static Set by **Create > Static Set > Create As Static Set** or by **RMC** (*above the selected project items*) > **Create As** > **Create As Static Set**. The selected items are now added to the newly created Static Set.

When Sets have been created, you can add additional project items to these sets by first selecting the project items that need to be added. Next, you go to **Create > Static Set > Add to Static Set** or by **RMC** (*above the selected project item(s)*) > **Add to Static Set**.

An overview of all members in a Static Sets is shown in *Navigation View* > *Sets* > *Static Sets* > *Name of the Set*. When you want to remove items from the set, select the project item and delete it. The project item will only be removed from the set and is not deleted in the project itself.

Grouping Data with Dynamic Sets

The dynamic set is based on an underlying query that retrieves project items from your project. The focus of the search determines the strength of the dynamic set. NVivo provides two types of searches to build dynamic sets: an intermediate and an advanced search type. We give an example of both types.

For the Mac user

- Dynamic Sets are not yet available in the MAC version.

Surprisingly, there is no "basic" or "advanced" definition of a search for a dynamic set. The least complex definition is called the "intermediate" search definition. This definition is shown when you create a new dynamic set by **Create > Dynamic Set**. First, you name the dynamic set in the tab *General*. In the tab *Search Criteria*, the actual search is defined. The option *Look for* allows you to specify which type of project you want to group with the dynamic set. Nearly all types of project items are shown in the drop-down menu. Underneath this option, two tabs refer to either the intermediate or the advanced search definition. In Fig. 11.6, we give the example of the dynamic set *Communities further from Beaufort* (available in *Navigation View* > Sets > Dynamic Sets). Selections are made in this window from left to right. You first select the type of selection you want to define and subsequently define the search with details on the right-hand side. The example in Fig. 11.6 searches for cases coded with a particular attribute value (see Chap. 9 for more information on attributes and values). Therefore, the left-hand option to be selected is *Classified items where*. The first option to be selected next is the actual attribute the selection is made on. In the example, a selection was made from the Case Classification *Place* and the attribute chosen is labelled *Driving time to Beaufort*. This attribute is defined as an integer type. The dynamic set needs to contain cases with a minimum distance of 30 min from Beaufort. Therefore, the second option to be selected is the actual driving time. Here, the operator *≥ value* was used to identify driving distances further than (or equal to) *24* min. When you click on the dynamic set (*Navigation View* > **Sets > Dynamic Sets > Communities further from Beaufort**), you will see a list of 7 items with cases *Atlantic, Cedar Isle*, etcetera.

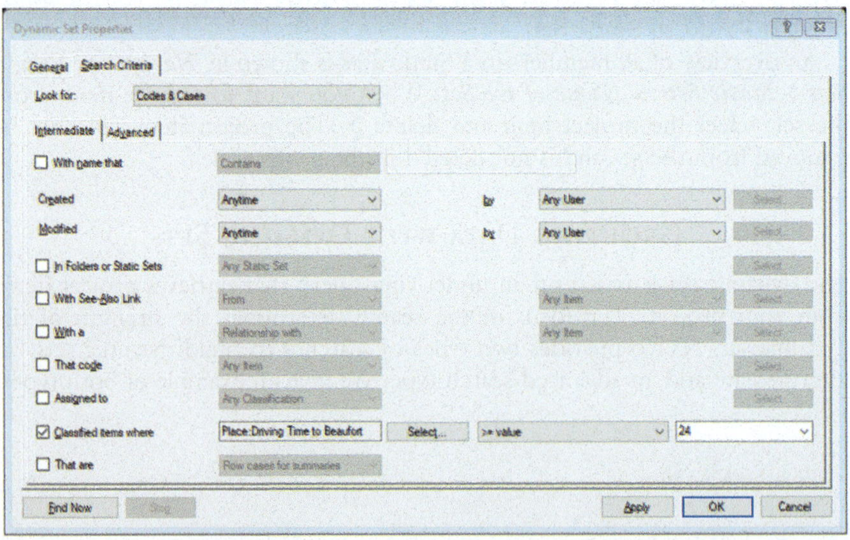

Fig. 11.6 Creating a dynamic set with an intermediate search

When you define a dynamic set with the Intermediate search criteria, you can only define one type of search string. You can combine different types of searches (e.g. searching for attribute values in combination with searching in a particular folder or set), but two queries of the same kind cannot be combined. In the Advanced tab, all search strings are defined individually, and multiple queries of the same type can be combined. Figure 12.7 gives an example of selecting two attribute values from the case classification Person. First, all males are selected (Gender = Male). This query is defined underneath the white area labelled *Find items that match all these criteria*. In the option *Interaction*, you choose the *attribute*. When selected, NVivo shows the window *Select Project items* and asks you to select an attribute from the case classifications available in the project. When you select Gender, the right-hand side of the query allows you to select the actual value (Female or Male) and an operator (in this example, *equals to*). Click on *add to list* to copy this first part of the query to the white box above and start creating the second query. This query is similar but states that the attribute *Generations Down East* is equal to *3 or more*. When you add this query to the list, see the example in Fig. 11.7. This Dynamic Set is saved in the Navigation View > Sets > Dynamic Sets > 3 + Generations Down East—Men in the sample project. In total, 35 items match both search criteria and are selected into the dynamic group.

With dynamic sets, we have an elaborate tool that allows us to create complex combinations of criteria across different project items. The selected results can easily be used to limit query results to a subsection of your data. Storing the selection criteria in these dynamic sets allows you to reuse the same selection repeatedly in different queries.

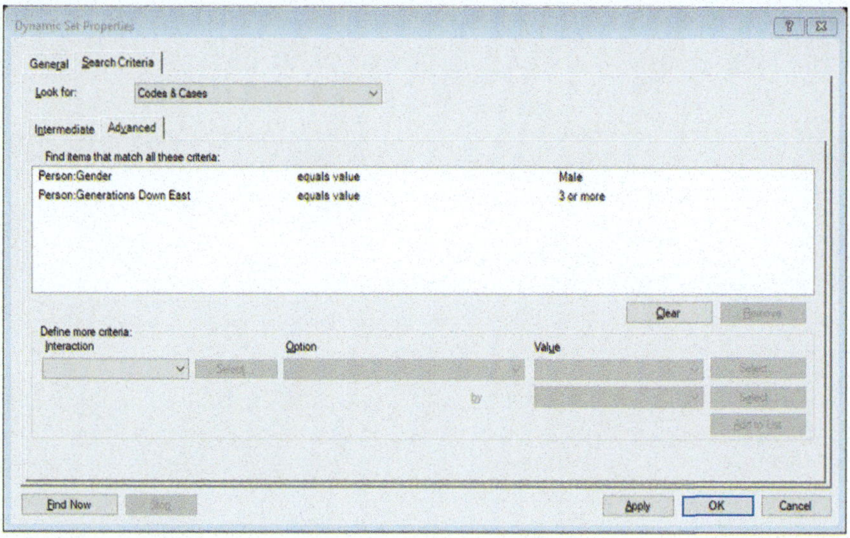

Fig. 11.7 Creating a dynamic set with an advanced search

CHAPTER 12

Querying (Coded) Data

Key messages in this chapter

- Queries help you to search for coded and non-coded material in your project.
- NVivo provides eight different queries that each allow to search in a different way.
- The Matrix Coding Query is a crucial analytical tool in NVivo.
- Queries can not only search for material, they also help you to code your material in specific situations.

Some Background on Querying

Queries are an important tool in NVivo. As explained earlier (see Section "Code and Retrieve as the Fundamental Principle" in Chap. 3), NVivo is a code-and-retrieve program whereby coded material is retrieved using queries to get a systematic overview of a specific topic in your data. These queries can be straightforward searches that retrieve coded information, but NVivo also allows the definition of very complex search functions that try to discover deeper connections in the data. The preparation and execution of queries is, therefore, a process that is directly connected to the coding process. For example, you can request information based on the codes on which the references are coded. Or you can request information that is coded simultaneously on several codes.

D. Mortelmans, *Doing Qualitative Data Analysis with NVivo*, Springer Texts in Social Sciences, https://doi.org/10.1007/978-3-031-66014-6_12

However, building and executing queries in NVivo is not done with a single uniform query tool. NVivo has eight different types of queries that each perform a different type of search in your project:

Coding query	Searches for coded material
Text search query	Searches for a text string in your data
Compound query	Combines a text search query with a coding query
Matrix coding query	Combines two lists of project items in a tabular form
Crosstab query	Combines a list of items with two attribute values
Word frequency query	Counts words in your (text) data
Group query	Finds associated items in your data
Coding comparison query	Calculates statistics for quantitative coding comparison

In what follows, we will introduce you to the different possibilities each of these queries offers. Before we go into the details of the queries themselves, we first discuss some common elements of the query (detail) window (Section "Dissecting the Query Window"), and we give a simple example to illustrate the work with queries (Section "Making Your First (Coding) Query").

Dissecting the Query Window

Each of the eight queries in NVivo can be defined by selecting the query in the menu **Explore**. Three queries are shown on the main menu: Text Search, Word Frequency and Matrix Coding Query. The other five query icons are hidden under the icon **Explore > Queries**. In this paragraph, we do not discuss the actual definition of the search. We first go into the layout of the query definition in the *Detail View* window.

Figure 12.1 shows the general layout of a query definition window. We use the Coding Query as an example in this figure, but all queries' layouts have a similar layout. The window is horizontally divided in two. The upper part of the window is the ***Query definition area***. In this area, you will define what to search for in your project. In the lower part of the window, we find the ***Results area*** where NVivo will show your search results once submitted. The queries are interactive tools. This means that when you run a query, the definition window remains in its place, and the results are added (or changed) in the results section. When you evaluate the results and want to change something to the search, you can easily change your query in the upper part of the window, rerun the query and re-evaluate the (modified) results.

Figure 12.2 shows some recurring buttons and options in the Query definition area. In several queries, the same buttons are shown in the same place. This makes the definition of queries consistent across different types of queries. The main difference between queries is the middle part of the definition

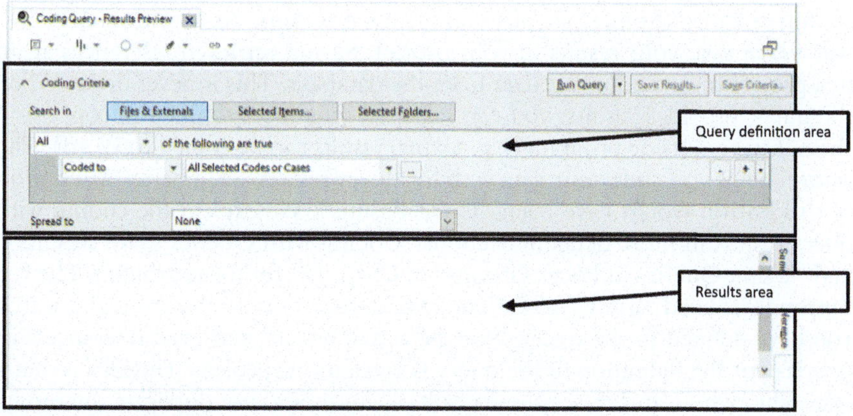

Fig. 12.1 Detail view window of (coding) query

window: this is where the actual search is defined, which depends on the query type you want to use.

Three recurring areas can be found in the definition section of a query. First (left top of the definition area), three buttons allow you to limit your search to a specific area in your project. The standard option here is *Files & Externals*. This button is selected in each new query and implies no limitations to the search: NVivo searches in all project items available to this query. The second button, *Selected Items,* lets you choose specific project items to search in a separate window *Select Project Items.* You can now choose all project items that can deliver results in your search. All items not selected will be excluded from the query results. The third button, *Selected Folders,* is similar, but now the window *Select Folders* will only allow you to limit the search to project items in specific folders. All folders not selected are excluded again from your search results.

The second recurring area (right top of the definition area) concerns the control of the query. The first (of three) buttons is called *Run Query*, which starts the search you have defined. As said earlier, you can constantly redefine a query and then re-run the query by clicking this button again. The second

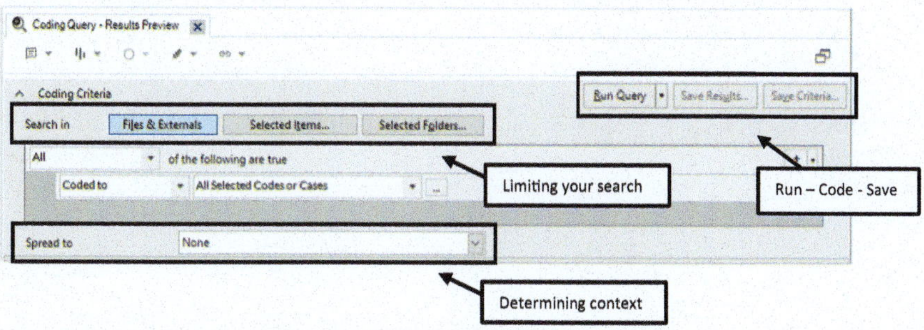

Fig. 12.2 Recurring tools in the query definition area

button is called (somewhat misleading) *Save Results*. As a principle, NVivo will ***never*** save your results in your project. Saving retrieved references again in a project means doubling data from the database. This is never done. When the query provides results, you can select all references found and copy-paste them into an outside program (e.g. textual results can be copied to Word). The button helps you code your data with the query results. So, a better description of this button would have been "Code Results". We explain the coding with query results in more detail in Sections "Coding with Queries (Save Results)" and "Using Queries in Focus Groups" in Chap. 15 (for focus groups). The last button is labelled *Save Criteria*. The criteria this button refers to is the search you have defined in the query. Not the actual results (retrieved references) of queries but the definition of the query is saved in the project. The idea is that a researcher who wants to see these results again should not duplicate the results in the project but just rerun the query to get the results again in the Results area. When clicking the *Save Criteria* button, you need to give the query a name, which is saved in *Navigation View* > **Queries** > **Query Criteria**. Each query type has its own icon referring to the type of search it can contain (see the icons in the list of queries earlier).

The third recurring element is the option *Spread to* at the bottom of the definition area. This option consists of a dropdown menu with six options. The options help expand the scope of the references NVivo shows when presenting references in the results area. The standard option is always *None*, meaning the results are given as they are found. This means that when you have coded on a word basis, NVivo shows only words when asked to give the results of that code. You can control the context shown in the results with the option *Spread To*. The following options are available to change the context of your results:

Coding reference	When using the text search query to search for a specific term, this option will show the surrounding reference.
Narrow context	This option shows five words before and after the result that is found. For sound and video, this is 5 seconds before and after. For pictures, it is defined as 5 per cent around the selected area.
Broad context	It is similar to the narrow context, but for text, it now presents the surrounding paragraph. For sound and video, the selection is expanded with 20 seconds. For pictures, the area presented is 20 per cent bigger.

(continued)

(continued)

Custom context	This option shows a separate window where you can choose exactly how you want to expand the context in terms of words, seconds, or percentages.
Entire file	This option shows the entire source as a result of your query.

For the Mac user

- The "spread to" option is not shown in the query definition in MAC. When you run a query and you RMC on the results, the option "show context" is shown in the context menu.

Not only in the Query definition area, we find recurring elements. Also, some specific elements are available in the Results area, independent of the query type (in most cases). The central part of the results area is the left white area where the results are shown. When a query has not yet run, there is just a white rectangular area. After running a query, the results are shown here. When the results consist of text, the references are immediately shown. When other output types are found, NVivo shows a Summary of the findings as a list (of files where results are found). In Fig. 12.3, we detail how these references are presented in the results area.

First, NVivo shows a link to the project item where the reference was found (as a blue hyperlink to the project item). The project item is highlighted, and when you click on the name, the item is opened in a new tab in the *Detail View* window. NVivo also mentions the total number of references found in that particular project item and shows a percentage *Coverage*. This percentage is later also shown next to the individual references. It presents a statistical figure on the space the reference(s) take. For text, the percentage refers to

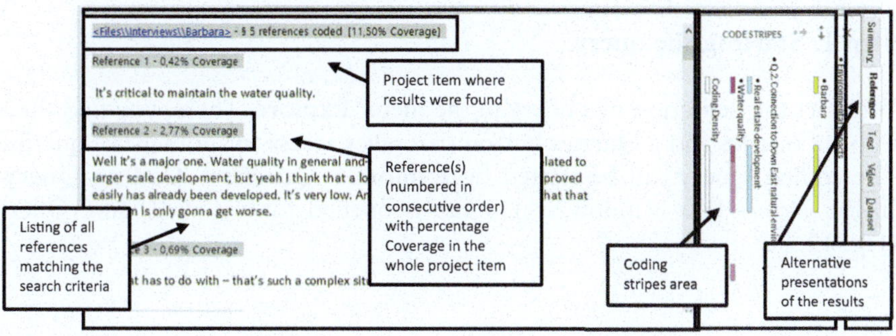

Fig. 12.3 Recurring tools in the query results area

the number of characters this reference takes in the total number of characters (also applies to PDFs and datasets). The percentage for audio and video files refers to the time this reference takes in the file. For pictures, the percentage is a percentage of the total amount of pixels in the file. It is important to note that the Coverage is a mere number and does not say anything about the importance of the reference for your analysis. Taking up more space (higher Coverage) does not imply having a more significant weight in your analysis. Some people describe with lots of words things that are of little importance to your analysis; others sometimes describe something precious for your analysis in a couple of words.

Within each file, all references are shown in a numbered list following the order in which they appear in the file. Again, for each reference, a Coverage is calculated. The results can be selected and copied to be exported to other programs. For example, textual references can be copied and pasted in your publication. An interesting feature in the results area is that you can activate your coding stripes and highlighting in this window. Each query window also has a small ribbon where the icons for both tools can be found. If the Coding Stripes take up too much space on your window, switch them off through the icon in the small ribbon.

A last recurring element in the results area is the tabs at the right-hand side of the area. They are difficult to spot, but you will see several tabs at a 90° angle. As indicated before, the tab References is selected when you have textual results from your query. When other types of results are shown, NVivo shows the *Summary* tab. Different tabs are shown on the right-hand side depending on the query or the results. In the example in Fig. 12.3, we also see tabs for Text, Video and Dataset (referring to results from these data types). We will present more tabs in the results area when discussing specific queries.

Making Your First (Coding) Query

This paragraph shows a stepwise example of how you create, run, adapt and save a query. We take the Coding Query to create the example, as this query is a basic query that many researchers will use in their analyses.

Step 1. Starting the query

You start the definition of a query in the menu **Explore**. There, you find three icons of queries and a Queries button that gives access to another five queries. The coding Query can be started by **Explore > Queries > Coding Query**. In the *Detail View* window, a new tab is opened for a new Coding Query (Fig. 12.4).

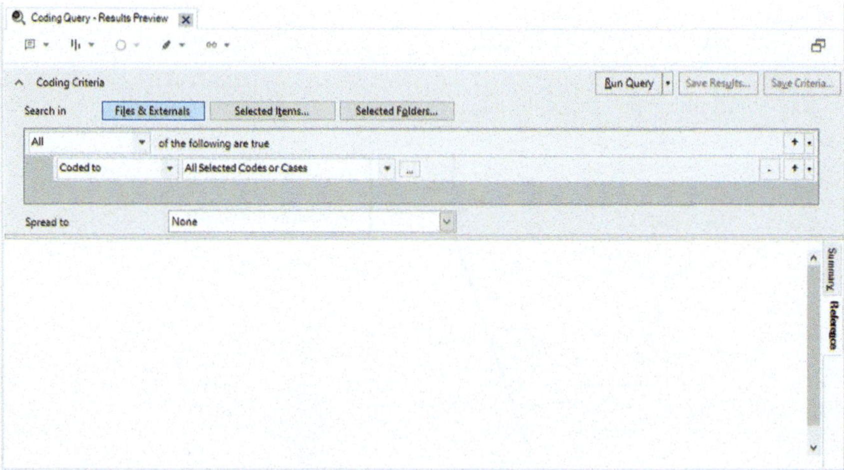

Fig. 12.4 Start screen for a new (coding) query

Step 2. Defining the initial query

In the first step, we want to search for all fragments coded with the code "water quality". To select this code, locate the three points in the middle of the definitional area. It is a tiny button after the two options *Coded to* and *All selected Codes or Cases*. When you click on this button, the Select Project Items window appears, and you can select the code "*Water Quality*" from **Codes > Natural Environment** (see Fig. 12.5).

When a code is selected, the name of the (first selected) code is now shown in the definitional area. This selection is the minimum requirement to run or save the query criteria. But before we go to step 3, we have also changed the context of the results by changing the *Spread To* option from *None* to *Broad Context*.

Step 3. Running the query and evaluating the results

To run a query, press the *Run Query* button, and NVivo executes the search across the data. In the results window, references that match the search are shown now (see Fig. 12.6). In the example in Fig. 12.6, we have activated the Coding Stripes and the highlighting for the Water Quality code. As such, you can see the results (highlighted in yellow) within their (broad) context. The References tab shows all references on the right side of the results window. Other tabs are available to focus on specific types of data (e.g. results from *Video*).

To evaluate the results, you can scroll down in the results area and review the references found with your initial search.

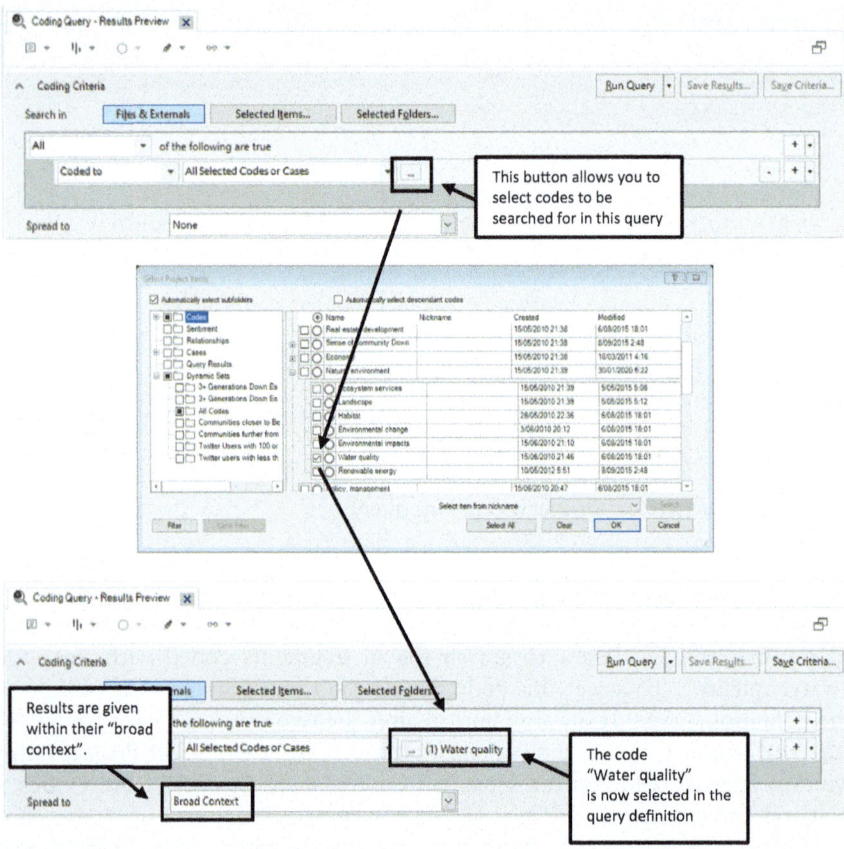

Fig. 12.5 Selecting a code to be searched in the coding query

Step 4. Adapting the query and rerun it

The queries in NVivo are interactive. That means we can adapt the first query we ran in the previous steps and re-run it again. For example, imagine that you go through the results and decide that you want to see the results from the code "Water Quality" and *the code "Renewable Energy"*. To accomplish this, you click on the select button (the tiny button with the three points) and select the code *Renewable Energy*" as well. You now have two codes selected (see Fig. 12.7). It is important to look at the option (at the same line) "*All Selected Codes or Cases*". This option requires all codes that are selected to be present in the *same* reference to be found. In the Boolean logic, this option refers to the keyword AND. But when we select the code "Water Quality" and "Renewable Energy", we did not want such a strict condition. We instead want all references where either of these codes is present. Therefore, we change the option to "*Any Selected Code or Case*". In Boolean logic, this option refers to

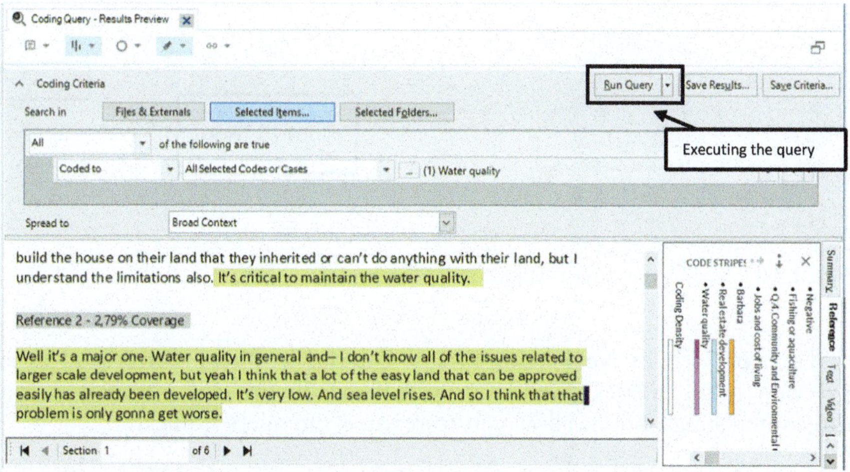

Fig. 12.6 Running the coding query

OR. To update the results, click the *Run Query* button again and evaluate the new results.

Step 5. Saving the (final) query definition

You can make some rather complex queries by finetuning the query criteria. If you are satisfied with the definition of a query, you might want to save your query-building work; that is what the button *Save Criteria* is for. This button shows a window where you must give the query a name. Afterwards, your query is saved as a project item in *Navigation View* > **Queries** > **Query Criteria**. You can now close the query tab. NVivo will warn you that any query preview results will be discarded. But when your definition has been changed, you can always return to the query definition and rerun it to get the results (Fig. 12.8).

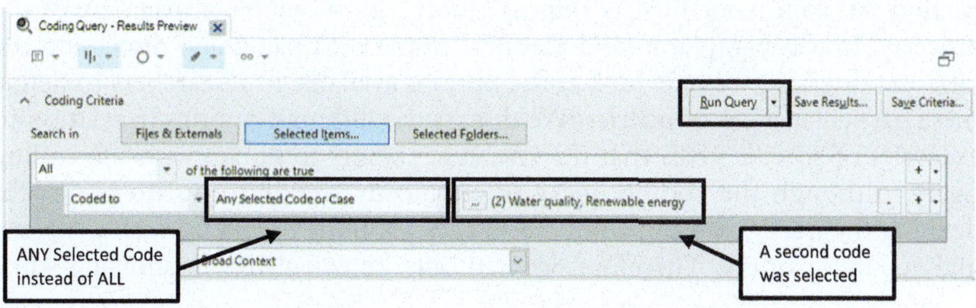

Fig. 12.7 Adapting and rerunning the query

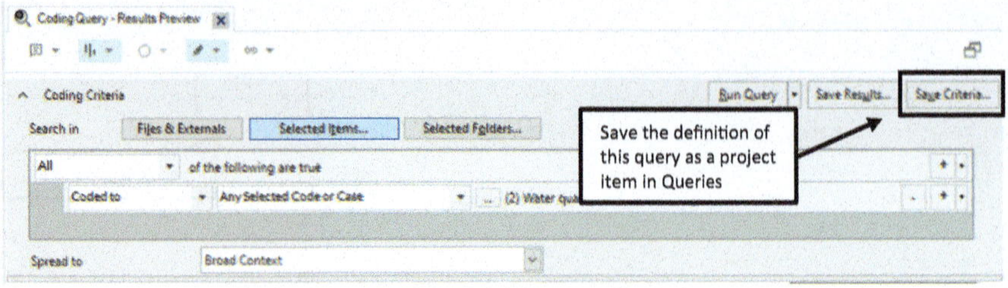

Fig. 12.8 Saving the query definition to the project

The *List View* in *Navigation View* > **Queries** > **Query Criteria** gives an overview of all queries saved as project items. In the RMC context menu, three interesting options allow you to reaccess the queries quickly:

RMC above the name of a query	*Effect of the option*
Run Query	Opens the query definition tab in the *Detail View* window and immediately runs the selected query, restoring the results in the results area.
Open Query	Opens the query definition tab in the *Detail View* window but does not run the selected query. You can either adapt the query definition first or run the query from the definition area.
Query Properties	Opens the properties window of this query, allowing you to change the name of the query and its description. The query definition cannot be changed, nor can the query be run from here.

The Coding Query (Some More Details)

In this paragraph, we elaborate on defining a Coding query. We will no longer go through the whole cycle of defining, running and saving the query (see Section "Making your First (Coding) Query" for a complete overview of the process). In this paragraph (and identical to the next paragraph discussing the different queries), we only look at the options available to researchers to define these particular types of queries. We discuss the different options in relation to examples of specific goals that the researcher wants to obtain with the coding query. Although the coding query offers mostly straightforward options for defining the query, you can combine all these options to end up with a surprisingly complex query. Therefore, we can only be exhaustive in showing some possibilities. Still, we are confident that this book's reader will grasp the available options and think of valuable combinations for their specific research. For the examples, we mainly explain the examples that are provided in the sample project. In the following screenshots, we will also focus on the definitional area, only showing how the options are implemented on your screen.

Example 1. Using attribute values

In the previous example, we only selected (content) coded material in our query. As working with attributes is also a form of coding, the coding query allows you to select attribute values to define the coding query. The first standard option is that *All selected Codes or Cases* need to be changed in *Any Case Where*. When clicking the selection button (the tiny button with the three points), you can now select attributes from your Case Classifications. Selecting attributes from File Classifications is impossible in a Coding Query (see Section "Choosing the Correct Type of Classification" in Chap. 9 for more information). When you have selected an attribute, NVivo shows you a drop-down menu with ten operators from which "*Equals value*" is selected. The ten operators give you various choices to select (or not select) a particular attribute value. The value is added in the last option. Here, the drop-down menu shows you all values that are (currently) available for that attribute. In the example in Fig. 12.9, we chose the attribute "*Person: Age Group*" and required the value to be equal to the age group "*30–39*".

Example 2. Using attribute values in combination with codes (using multiple rows)

In the second example, we build upon the previous one but add a second criterion to the age requirement. We want to know how respondents under 40 look at the fishing industry's decline. To do so, you click on the + symbol at the right side of the rows starting with "All" or starting with "Coded at". You choose "*Add row*" from the drop-down menu and see a new row starting with "Coded at" appear underneath the existing one (see Fig. 12.10). In that second row, we now select the codes ("*Due to regulations*", etc.) that we want to see in our results, and we opt for "*Any of the selected codes or cases*" to use the OR-operator (the All-operator would be too strict in this case and produce no results).

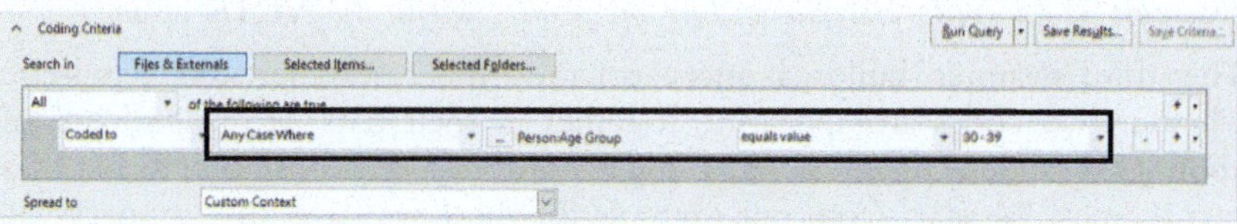

Fig. 12.9 Selecting an attribute value in a coding query

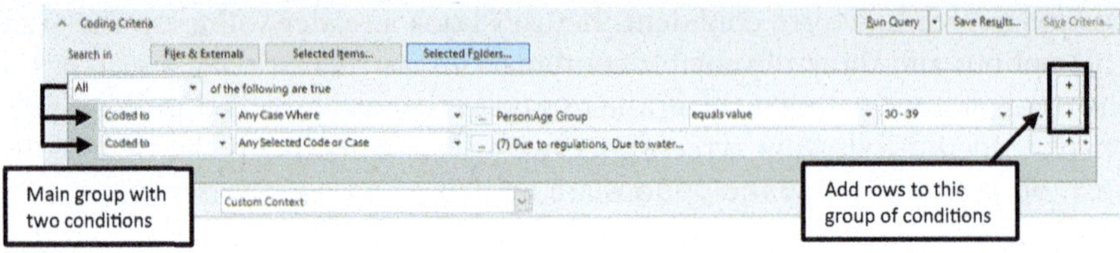

Fig. 12.10 Adding rows with extra conditions to a group

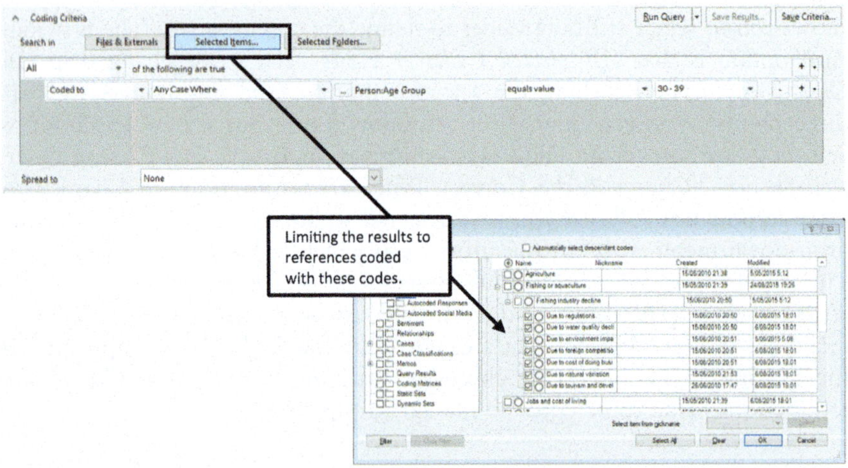

Fig. 12.11 Limiting a search with the Selected Items button

We now have one group of conditions with two rows. At the top of the group, the option "*All*" (*of the following are true*) indicates that both conditions need to be true simultaneously for results to be selected. So results will be coded at one of the selected contents codes, AND the references will all be from respondents in the age group 30–39 years old.

Side note: there is an alternative way of getting the same results without adding a second row. When you start with the query from Example 1, you can also "limit" the result by clicking on the button "Selected items" and selecting the codes from that window (see Fig. 12.11).

Example 3. Layering criteria within one group (using the NEAR option)

This third example builds a query whereby the researcher asks themselves whether two codes are appearing together in their data. If respondents talk about two (coded) phenomena, it might point to a relationship between these two themes. We start from our example 2, where we have built a query with two conditions: aged 30-90 and coded at codes under the parent code "*Fishing industry decline*". We now want to know whether these codes were also coded with "Negative" under the parent code "Attitude" to get the outspoken negative references to reasons for the decline in the fishing industry. To link the second condition to the first, click on the drop-down menu (the triangle icon at the right-hand side of the condition) and choose the option *Near*. The new condition is added as two intended rows underneath the existing condition. The first intended row describes the link and presents the suggestion of "*Overlapping*", which indicates that the subsequent condition needs to be in a reference that (partly) overlaps the results of the original condition. The second intended row allows you to select the codes NVivo needs to search in this overlapping condition (Fig. 12.12).

The advantage of this *Near*-option is that results do not need to be in precisely the same reference to be found. When you define multiple rows, the assumption is made that these conditions apply to the same reference to be found. With the Near-linkage, you can also search for results across non-identical references (the range of this option is 20 words of content).

Example 4. Using a user's work as a criterion

The fourth example describes a similar variation as in example 3. Again, we start from the search we defined in example 2. We know that the sample project from QSR has been constructed by three users: Effie, Henry and Wanda. We now want to know whether the codes we have identified were used by Effie or Henry. Again, we click on the drop-down menu at the right-hand side of the condition (click on the triangle symbol). We now choose the option *Coded by any*. Only one intended row appears, and the *Users* window lets you choose the users you want to see the coding work from. We select *Effie* and *Henry* from the list. These are shown in the new criterion in Fig. 12.13.

For the Mac user

- In the Mac version, using User's information as a criterion in queries is not yet possible.

Example 5. Using multiple groups

In example 2, we showed how two (or even more) conditions could be grouped and linked by requiring them *All* to be true (AND-linkage) or *Any* to be true (OR linkage). These two conditions were added to the main group of the Coding Query (see Fig. 12.10). This main group is always present and links all individual conditions and groups created underneath the main group. Sometimes, we need multiple sub-groups to search for specific patterns in our data. In this last example, we ask ourselves for references where men are talking negatively about the water quality and women have opposing opinions: we want references where they are talking positive about the water quality.

This is done in the following steps:

Step 1: Create a new Coding Query and remove the first condition automatically presented

When creating a new Coding Query, NVivo assumes you want to define a condition immediately. In this example, we do not want independent conditions outside sub-groups, so we click on the minus (−) symbol to remove this condition. Only the main group is left (see Fig. 12.14).

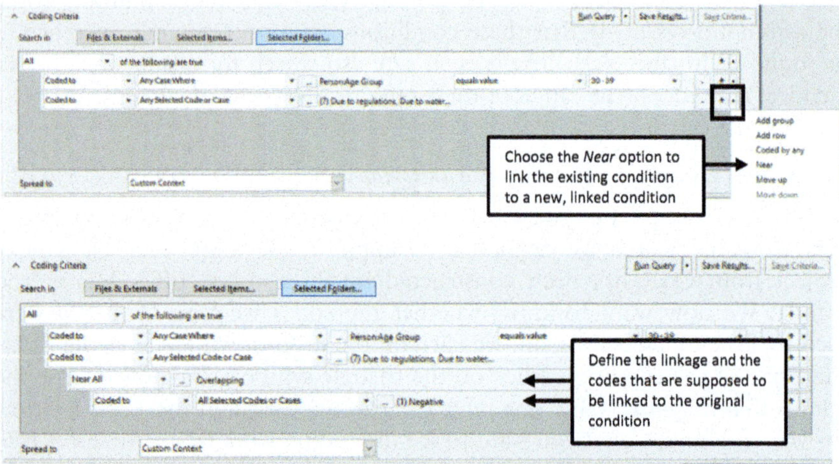

Fig. 12.12 Using the NEAR option to link two conditions

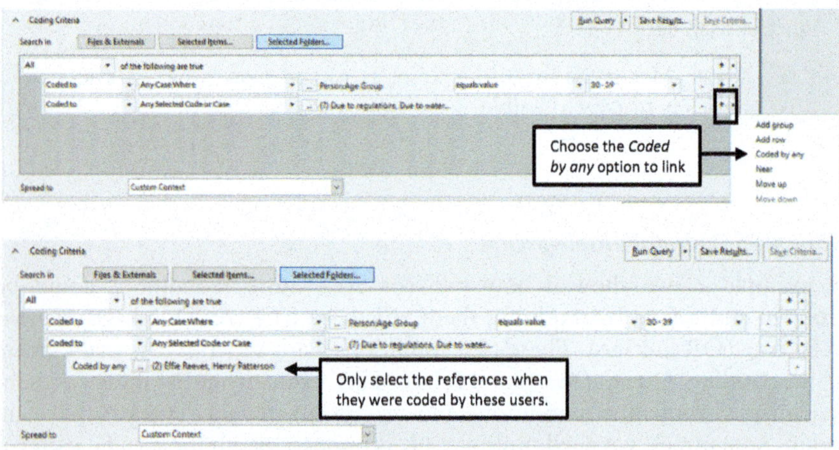

Fig. 12.13 Using information from users in a coding query

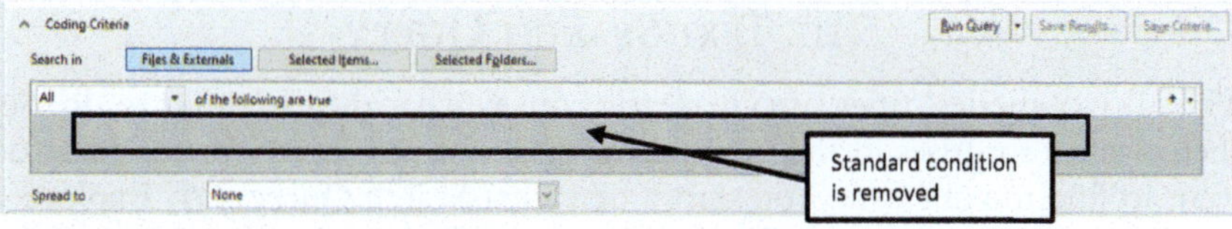

Fig. 12.14. Removing all standard conditions from a new coding query

Step 2: Create new (empty) groups

Next, we click the drop-down menu under the plus (+) icon and choose Add group. Repeat that to create a second group (Fig. 12.15). Note how NVivo automatically adds one condition to each new group. As we want to define these groups in the next step, these conditions will be the starting points in our next step.

Step 3: Define each of the new groups

In the final step, we defined each group separately to cover the search criteria we defined earlier. We used a combination of all previous examples to define both groups. For more details on the definitions, we refer to the examples above. We use the option *Any* as the link between our two groups. We want men talking negatively OR women talking positively. So, the conditions in either group must be true to be selected as a result. In the subgroups, we selected the option *All* as all the conditions within each group must be fulfilled to be selected as a result (Fig. 12.16).

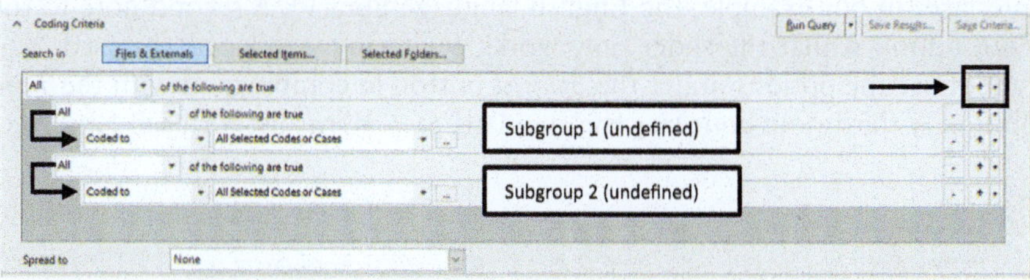

Fig. 12.15 Adding multiple groups in a coding query

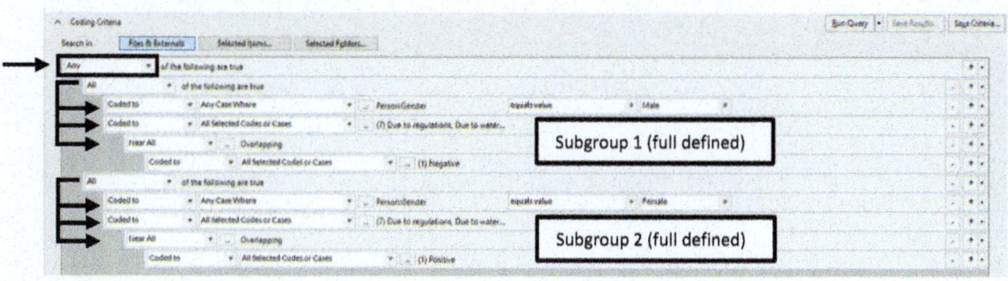

Fig. 12.16 Defining multiple groups in a coding query

The Text Search Query

The NVivo queries' titles usually clearly convey what they are built to do. So, there are no surprises in the *Text Search Query*: it will search for the text you want to find in your data. You start a new Text Search Query with **Explore > Text Search**. In the *Detail View* window, a new tab is opened for your Text Search Query (Fig. 12.17). The option *Search for* above the white central rectangular is the heart of the Query definition area. As the rectangular says in light grey: "*Enter the text to search for*" you can type your search term(s) in that area and search for its occurrences across the project. Again, the option *Spread To* will determine whether the results will only contain your search terms as you defined them or whether you will get the context around the search terms as well (e.g. by changing this option to *Broad Context)*.

Two options help refine the standard search for terms: the *Special* button and the slider in the *Find* option. The button *Special* contains nine options that help you to combine different search terms in a specific way (Table 12.1).

Two more options are available that are not shown under the *Special* button (Table 12.2):

At the right of the Query definition area, you find an option *Find* with a slider positioned standard at the option *Exact matches* (e.g. "talk"). The following four options determine which variations of a single search term can be found. For each option, NVivo provides an example of how the variations are considered (see Fig. 12.17). A first important note is that this slider only works with the official languages NVivo is available: English, French, German, Spanish, Japanese, Portuguese, or Simplified Chinese. No language library is available for other languages, and the slider will use the language library of the interface (in our example, the English language library). A second note to the *Find* option is that the slider only works when one search term is used. You cannot use the options under the *Special* button in combination with the *Find* option. A third note concerns the standard *Stop Words* list for each language.

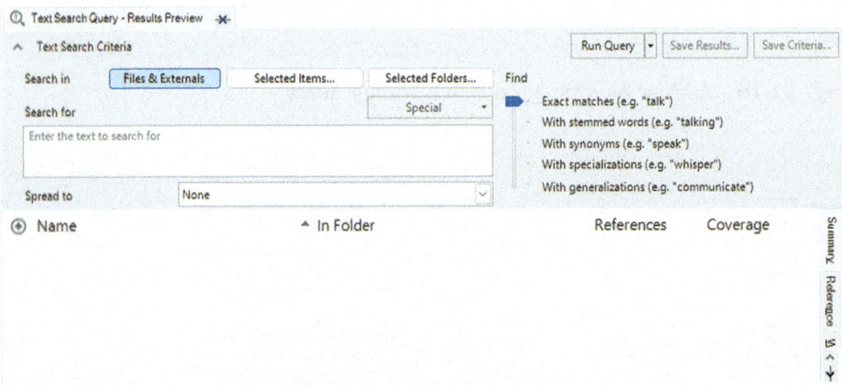

Fig. 12.17 Start screen for a new text search query

Table 12.1 Overview of options under the special button of the text search query

Option (under button special)	*Example*	*Explanation*
Wildcard—any one character (?)	Health?	Finds healthy, healths
Wildcard—any character (*)	Health*	Finds healthy, healthful, healthiness, healthward, healthify
AND	Health AND poverty	Both words need to be found to give a result
OR	Health OR poverty	One of the words needs to be found
NOT	Health NOT poverty	Health is found where poverty is not found
Required (+)	+ health poverty	Health is required, BUT poverty is also found
Prohibit (−)	− poverty health	Health is found in places where poverty is not found
Fuzzy (~)	Health~	Find words with a similar spelling of health
Near (~ with a number)	"Health poverty" ~3	Health and poverty are maximally three words separated. The number of words (here: 3) can be chosen.

Table 12.2 Overview of extra options (not under the special button) in the text search query

Option	*Example*	*Explanation*
Round brackets ()	(health~) AND (poverty OR poor)	Two separate searches (health and poverty) are joined with the Boolean expression AND.
Double quotes	"Health insurance"	Searches for the exact term "health insurance" as it is put in the double quotes.

By default, each of the main languages has a built-in Stop Words list of words excluded from your text searches. To see the Stop Words list active for you, go to **File > Project Properties > General** and click the Stop Words button to see which words are on the list.

The Matrix Coding Query

The Matrix Coding Query is one of the most powerful analytical tools in NVivo. As we will show in this paragraph, its definition is relatively straightforward and uncomplicated, but the range of possible analyses is enormous.

Creating a Matrix Coding Query starts with **Explore > Matrix Coding Query**. The Query definition area is now divided into a definition area for the rows (left) and a definition area for the column (right) (Fig. 12.18).

You can select items or add attribute values when clicking on the + symbol below the definitional areas. When choosing the first option, NVivo shows the *Select Project Items* window, where you can select items to be shown in either rows or columns. When you opt for the second option, you can only select attributes and their values from either File or Case Classifications.

In Fig. 12.19, we have selected the codes under the parent code Economy in the rows and added two attributes from the Case Classification Person to the columns. When running the query, the results section shows a table with these codes in the rows and the attribute values in the columns. The cells are all numbers, not texts, as with the previous two queries. What do these numbers represent? To find that answer, look at the extra **Matrix** menu now (Fig. 12.20). The standard meaning of the numbers is *Coding references*. But as shown in the figure, you can easily change this representation to have a different meaning. Is this important? No, the number shown in the cells is not that important. Only a value of 0 shows that there is no data in your project at the intersection of this row and column. All other numbers show you there is some piece of data there. It is important that you double-click on the cell actually to see the data behind the number.

We want to present two final options: the cell shading and the transpose possibility. First, you can give your table a background colour based on the results shown. With **RMC** (above the matrix) **> Cell Shading > Color scheme,** you can choose which colours you want to give to your table (see Fig. 12.21). With **RMC** (above the matrix) **> Transpose,** you change the rows and columns of place. This makes the table easier to interpret in some cases.

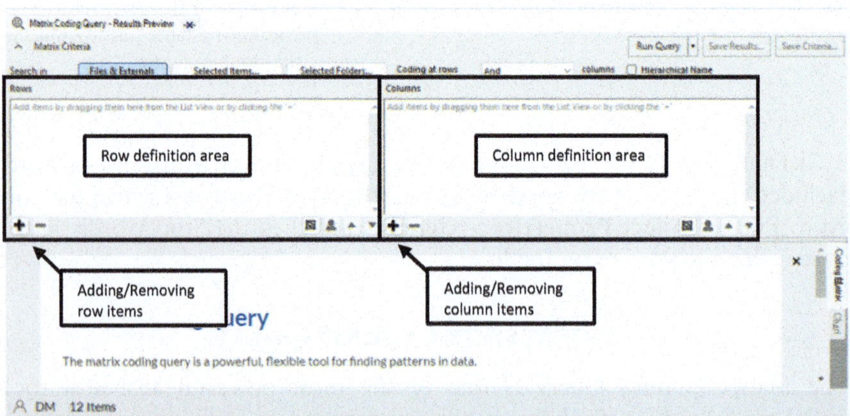

Fig. 12.18 Start screen for a new matrix coding query

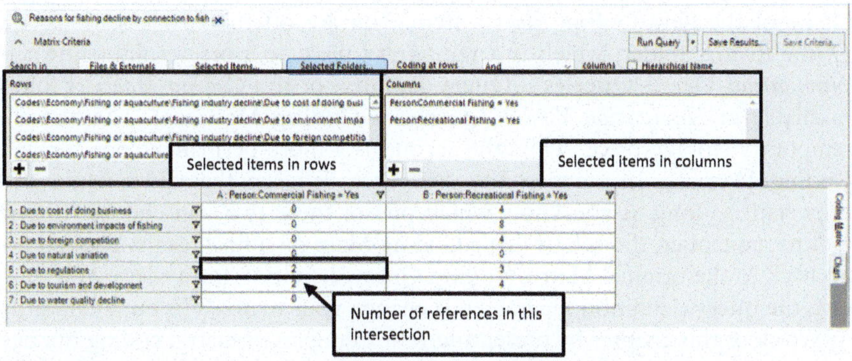

Fig. 12.19 Selections made and results shown in the matrix coding query

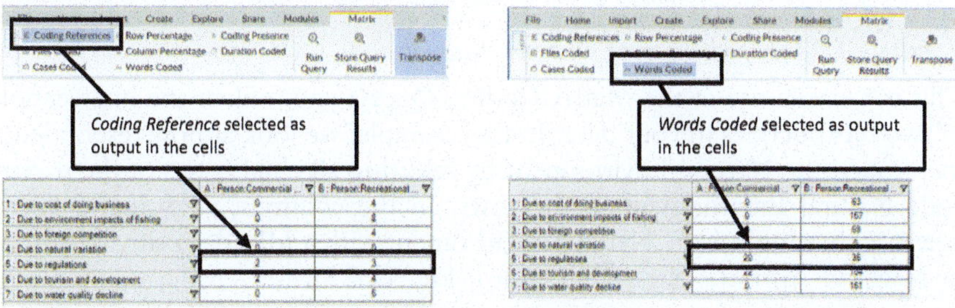

Fig. 12.20 Different output selections in the cells of a matrix coding query

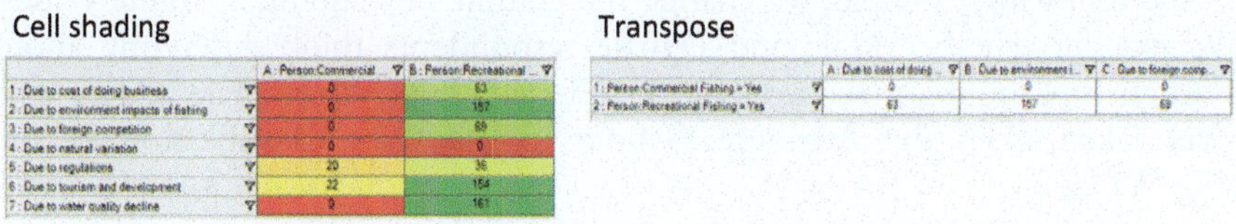

Fig. 12.21 Cell shading and transposing the result of a matrix coding query

Further in this chapter (see Section "Coding with Queries (Save Results)"), we will explain how you use the *Save Results* button to code with queries. In the Matrix Coding Query, however, the button *Save Results* has a different effect, so we discuss its use in this paragraph. Earlier, we explained that the results of queries are never copied in the project as this would mean data duplication, which is never done in a database. The same applies to the Matrix Coding Query. When you click the *Save Results* button, your results will not be duplicated. Instead, NVivo allows you to save the table definition as a Coding Matrix. This is a separate project item whereby the Matrix is saved and whereby you can easily reconstruct its results by opening this project item. When you click the *Save Results* button, you can choose the Location and the

Name of the new *Coding Matrix* item. You are offered two Locations: the *Query Results* folder, which is a temporary place to store results (shown in *Navigation View* > **Queries** > **Query Results**) or the *Coding Matrices* folder (available in *Navigation View* > **Queries** > **Coding Matrices**). The first is a temporary storage place, while the second is a more permanent one for final Coding Matrices. In practice, the difference between these two is not that important as long as you know which one of these two you choose. Important to remember, though, is that a Coding Matrix is a final result. You can not go back to the original Matrix Coding Query definition from there. You must save the query definition separately to change your matrix's layout afterwards.

We close this chapter by presenting four examples of matrix coding queries. We will shortly introduce the example and show you the screenshot of the final result. We hope these examples illustrate the potential of this query and may inspire you when analysing the data in your project.

Example 1. Only coding work in rows and columns

This is a classic use of the Matrix Coding Query: you look at the intersection of two sets of codes in your data. In this example, we look at how respondents mentioned the Natural Environment by selecting all child codes of the parent code Natural Environment to the rows. In the columns, we want to know whether they look at the Natural Environment with a Mixed, Positive, Neutral or Negative Attitude (the parent code) (Fig. 12.22).

Example 2. Using attribute values in the columns

In the following example, we change the columns by adding attribute values. We wonder whether Male and Female respondents think differently about Natural Environment (see example 1). The attribute values Male and Female were taken from the Attribute Gender in the Case Classification Person (Fig. 12.23).

Example 3. Using files or cases in the rows

The third example shows how a Matrix Coding Query can use other project items than codes. In this example, we put all cases of our Interview Participants in the rows, and we put the attribute values of the attribute Township

	A : Mixed	B : Negative	C : Neutral	D : Positive
1 : Ecosystem services	0	1	0	2
2 : Environmental change	2	10	0	2
3 : Environmental impacts	1	20	0	0
4 : Habitat	1	14	0	1
5 : Landscape	0	0	1	29
6 : Renewable energy	0	5	0	3
7 : Water quality	12	60	9	60

Fig. 12.22 Example 1 of the matrix coding query (resulting matrix)

	A : Person:Gender = Female	B : Person:Gender = Male
1 : Ecosystem services	5	3
2 : Environmental change	22	15
3 : Environmental impacts	2	8
4 : Habitat	11	3
5 : Landscape	17	17
6 : Renewable energy	0	0
7 : Water quality	51	73

Fig. 12.23 Example 2 of the matrix coding query (resulting matrix)

(also in Case Classification: Person) in the columns. As such, we get an overview of which respondents live in which township of the city we study. This overview can help to see whether we have sufficient regional diversity among our respondents (if we assume this is important for you as a researcher) (Fig. 12.24). Note with this example that the numbers in the cells have little meaning other than the total number of references coded in these files. For this example, changing the representation of the cells to *Files Coded* would make more sense.

	A : Person:Township = Straits	B : Person:Township = Harkers Island	C : Person:Township = Marshallberg
1 : Barbara	21	0	0
2 : Betty	1	0	0
3 : Charles	0	0	0
4 : Daniel	0	0	0
5 : Dorothy	0	0	0
6 : Helen	7	0	0
7 : James	0	0	13
8 : Ken	0	0	0
9 : Margaret	0	0	0
10 : Maria	0	0	0

Fig. 12.24 Example 3 of the matrix coding query (resulting matrix)

	A : Coding by EDR	B : Coding by HGP	C : Coding by WWS
1 : Ecosystem services	6	2	10
2 : Environmental change	27	38	39
3 : Environmental impacts	7	7	26
4 : Habitat	19	13	37
5 : Landscape	29	8	35
6 : Renewable energy	0	0	15
7 : Water quality	120	15	138

Fig. 12.25 Example 4 of the matrix coding query (resulting matrix)

Example 4. Inserting users in the columns

For the final example, we put the codes again underneath the parent code Natural Environment in the rows. But now we have inserted the three users who worked on the example project at NVivo in the columns: Effie (EDR), Henry (HGP) and Wanda (WWS). We consider this example as an alternative for the Coding Comparison Query (see Section "The Coding Comparison Query"). Instead of producing statistical indicators on coding work, qualitative researchers are better off discussing their coding work among colleagues. In this example, different coders can compare their coding work (provided it was done on the same material). The different numbers indicate a different use of the same codes. This is a starting point among different qualitative researchers to talk about the meaning of codes and their implementation to the material in the project. Such an approach is often more fruitful than relying on the statistical output, e.g. a Kappa indicator (Fig. 12.25).

For the Mac user

- Example 4 cannot be done in MAC as the MAC version of NVivo does not yet allow to add users to this query

The Crosstab Query

The *Crosstab Query* (**Explore > Queries > Crosstab**) is similar to the *Matrix Coding Query*. It allows users to make queries similar to what we presented in Example 2 of the previous paragraph (see Section "The Matrix Coding Query"). The Crosstab Query is designed to put codes in the rows and intersect them with attribute values in the columns. The major difference with the Matrix Coding Query is that this query allows you to cross-tabulate two (exact two) attribute values (either from one File Classification or from one Case Classification). In example 2 of the previous paragraph, we showed how we made a matrix with the child codes of Natural Environment in the rows

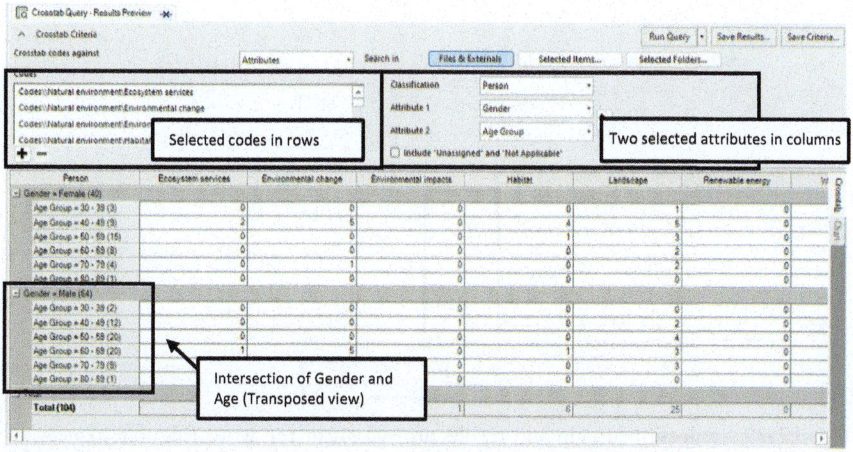

Fig. 12.26 Selections made and results shown in the cross-tab query

and the attribute values Male and Female in the columns. Suppose now that we were also interested in the age of our respondents. We could have also added the attribute Age Groups to the columns of our Matrix Coding Query. However, the Age Groups would have been added as separate columns in the matrix. They would not allow us to analyse old and young men and old and young women. This issue is resolved in the Cross-tab Query. In the columns, we can now put both Gender and Age Groups, and the resulting matrix will give the cross-tabulations of these two in the columns (Fig. 12.26). Note that in the screenshot of Fig. 12.26, we have transposed the table to show the intersection of Gender and Age more clearly. In the original output, the intersection was shown in the columns.

Other NVivo Queries

In the previous paragraphs, we discussed the four major queries in NVivo: coding, text search, matrix coding and cross-tab. But NVivo has eight different queries in total. In this paragraph, we give a short overview of the last four queries without exploring them in the same depth as the major ones because we assume they will be used much less than the queries we described earlier.

The Word Frequency Query

The Word Frequency Query (**Explore > Word Frequency**) counts words in your (textual) files. It comes with a minimal number of options. You can determine the amount of words that needs to be represented in the frequency table, and you can also determine their minimal length. Comparable to the Text Search Query, the *Grouping* option allows you to group multiple terms

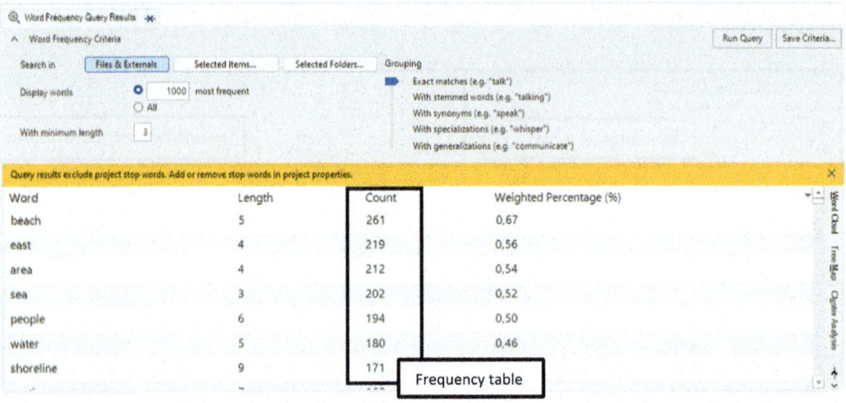

Fig. 12.27 Selections made and results are shown in the word frequency query

(for the installed and active language in this project). The result is shown in Fig. 12.27.

The Word Frequency Query has many alternative visual representations available, both from the tabs at the right side of the Query Results area and from the *Word Frequency Query* menu. Some examples in Fig. 12.28.

For the Mac user

- The tab Cluster Analysis is not yet available in MAC.

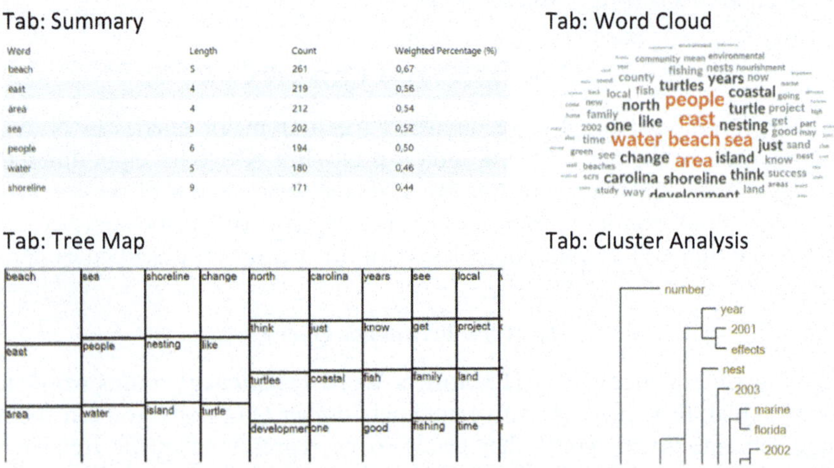

Fig. 12.28 Different visualisations in the word frequency query

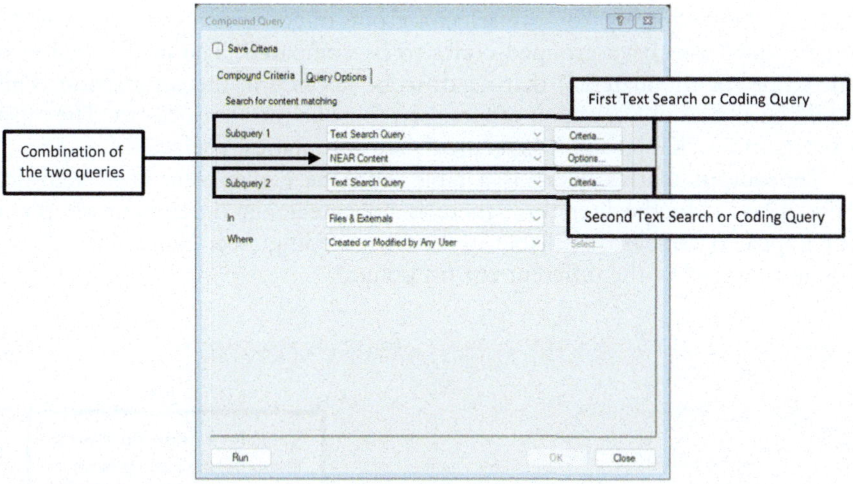

Fig. 12.29 Definition of the compound query

The Compound Query

The *Compound Query* ((**Explore > Queries > Compound**) allows users to make a combination of a Text Search Query and a Coding Query or a combination of two Text Search or two Coding Queries. We refer to the respective basic queries (see Sections "The Coding Query (Some More Details)" and "The Text Search Query") to define these types of queries. The two queries can be combined with the NEAR options discussed in Example 3 of Section "The Coding Query (Some More Details)" (Fig. 12.29).

> **For the Mac user**
>
> - The Compound Query is not available in MAC.

The Coding Comparison Query

The Coding Comparison Query (**Explore > Queries > Coding Comparison**) will make little sense in many qualitative analyses. Still, it might make sense to use NVivo for a quantitative content analysis. NVivo can be used for that kind of analysis in such a context. However, since NVivo allows sufficient flexibility to use the program for quantitative content analysis, this query can produce statistical output that compares quantitative coding work.

The definition of the query is shown in Fig. 12.30. The user must first define the user or users that are considered the first group to be compared with those who form the second group. Next, you define the coded material that needs to be compared. You can select a group of codes or a static or dynamic set in which you have grouped codes to be compared. The final selection is the scope (or the material) that needs to be selected in the comparison. This could be all files or a selection of files used to make the comparisons. The most used statistics (Kappa and %Agreement) are automatically selected.

The output of this query is a table with the earlier-mentioned statistics (Fig. 12.31). In the literature, there is no agreement on the cut-off point for Kappa to establish a reliable quantitative coding (see: Neuendorf, 2017, for an overview of the different cut-off points).

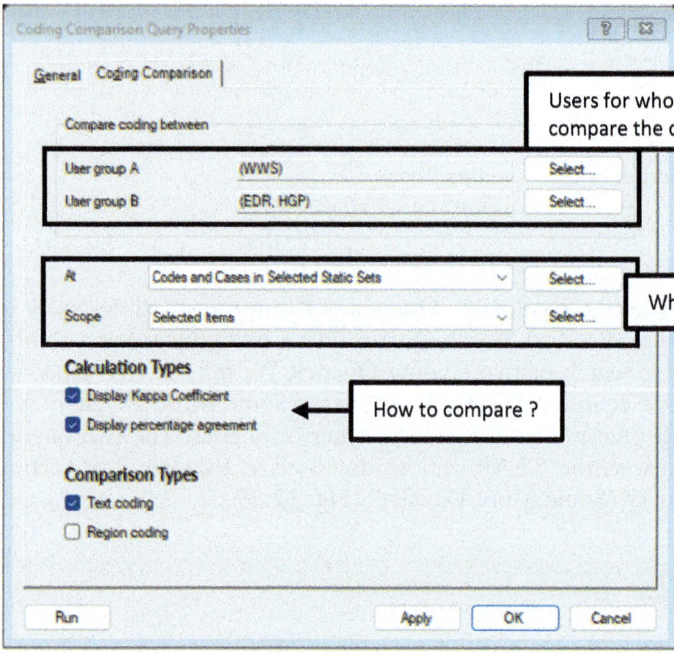

Fig. 12.30 Definition of the coding comparison query

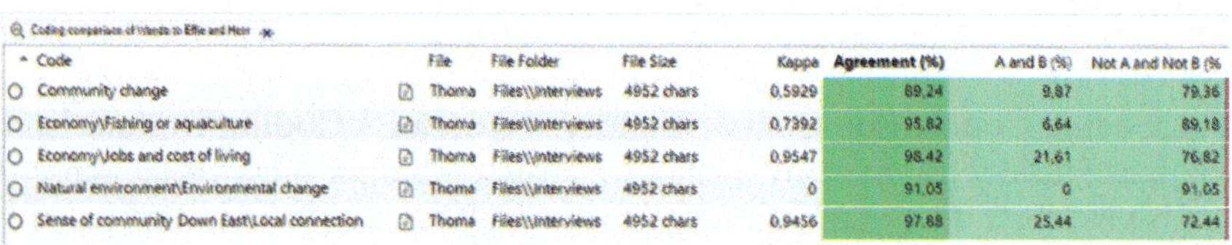

Code	File	File Folder	File Size	Kappa	Agreement (%)	A and B (%)	Not A and Not B (%
Community change	Thoma	Files\\Interviews	4952 chars	0,5929	89,24	9,87	79,36
Economy\Fishing or aquaculture	Thoma	Files\\Interviews	4952 chars	0,7392	95,82	6,64	89,18
Economy\Jobs and cost of living	Thoma	Files\\Interviews	4952 chars	0,9547	98,42	21,61	76,82
Natural environment\Environmental change	Thoma	Files\\Interviews	4952 chars	0	91,05	0	91,05
Sense of community Down East\Local connection	Thoma	Files\\Interviews	4952 chars	0,9456	97,88	25,44	72,44

Fig. 12.31 Output of the coding comparison query

The Group Query

A Group Query (**Explore > Queries > Group**) provides an overview of items that appear together in a project (literally: grouped). It is a query that gives descriptive overviews of your project's linked elements. For example, the Group Query lets you get an overview of all codes used in a particular interview. Or it can produce the reverse: which interviews have been coded with a particular code?

The basic choice you need to make is what type of project item you want to get summarised. In the option *Look* For, NVivo gives you seven choices of project items that can be summarised (Fig. 12.32). These items (to be summarised) are referred to as the *Range Items*. They are summarised within the *Scope Items* of your choice. We give an overview of the choices in Range and Scope items in Table 12.3.

We will only provide an overview of some output generated by the examples in Table 12.3. We limit ourselves to the first example (*Which codes were used to code the interview with Barbara?*) only. The Group Query produces a tabular overview (left panel in Fig. 12.33) and a visualisation of the results using a Connection map (right panel in Fig. 12.33). In the tabular output, the Scope Item was the interview of Barbara. That is presented as the main entry in the list. Since we only asked for that interview, only one Scope Item is shown.

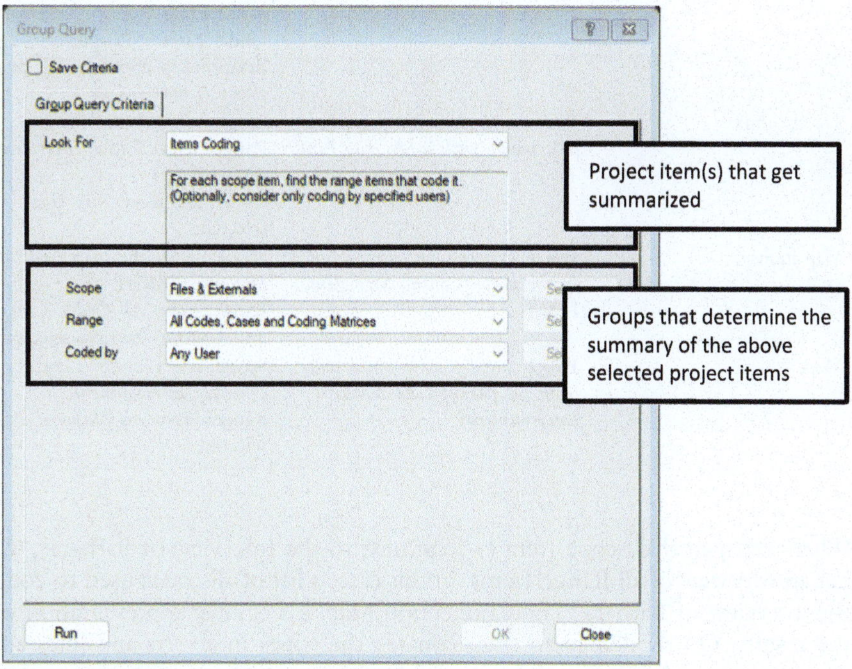

Fig. 12.32 Definition of the group query

Table 12.3 Overview of options in Look For in the Group query

Choice in Look For	*Example*	*Choices to make*
Items coding	*Which codes were used to code the interview with Barbara?*	Scope: select interview Barbara (under *Files / Interviews*) Range: select folder *Codes* Coded By: leave *Any User* selected
Items coded at	*Which interviews were coded with the child codes under the parent code Natural Environment?*	Scope: select all child codes of *Natural Environment* Range: select folder *Interviews* (under *Files*) Coded By: leave *Any User* selected
Items by attribute value	*Which interviews have highly educated participants?*	Scope: select attribute value *Completed graduate school* in Case Classification *Person* of Attribute *Educational Level* Range: select folder *Interviews* (under *Files*)
Relationships	*Which cases contain the relationship 'Is married to'? (Or: who is married to whom?)*	Scope: select folder *Interview participants* under Cases > People Range: select folder *Interview participants* Under cases > People Relationship criteria > Types: select *Is married to*
See also links	*Which interviews have referrals to a memo with a See Also Link?*	Scope: select folder *Memos* Range: select folder *Interviews* (under *Files*) Coded By: leave *Any User* selected
Map items	*Which project items are included in the map "Coding Structure FINAL."*	Scope: select the map *Coding Structure FINAL* Range: leave *All Codes, Cases and Coding Matrices* selected
Map	*Which maps contain child codes from the parent code Natural Environment?*	Scope: select all child codes of *Natural Environment* Range: leave *All Maps* selected

When we open this Scope Item (+ icon next to the Interview of Barbara), we get an overview of all Range Items, in this case, a list of all codes used to code the interview of Barbara. The Connection Map is a circular representation of the results. On the left-hand side, you see the (one) interview and lines are drawn to the codes on the right-hand side. Of course, this visualisation would be more informative if we had added more interviews. Then, a comparison of coding across interviews would have been possible.

For the Mac user

- The Group Query is not yet available in MAC.

Tab: Summary

Scope Item	In Folder	Finds	Created on
Barbara	Files\\Interviews	32	27/05/2010 13:03

Range Item	In Folder	Created on	Created by
Sense of community Down East\Local ide	Codes	28/05/2010 22:25	WWS
Sense of community Down East\Local co	Codes	15/05/2010 21:38	WWS
Real estate development	Codes	15/05/2010 21:38	WWS
Policy, management	Codes	15/06/2010 20:47	HGP
Natural environment\Water quality	Codes	15/06/2010 21:46	HGP
Natural environment\Landscape	Codes	15/05/2010 21:39	WWS
Natural environment\Habitat	Codes	28/05/2010 22:36	WWS
Natural environment\Environmental chan	Codes	3/06/2010 20:12	HGP
Natural environment\Ecosystem services	Codes	15/05/2010 21:39	WWS
Natural environment	Codes	15/05/2010 21:39	WWS
Infrastructure	Codes	15/05/2010 21:38	WWS
Economy\Jobs and cost of living	Codes	15/05/2010 21:39	WWS

Tab: Connection Map

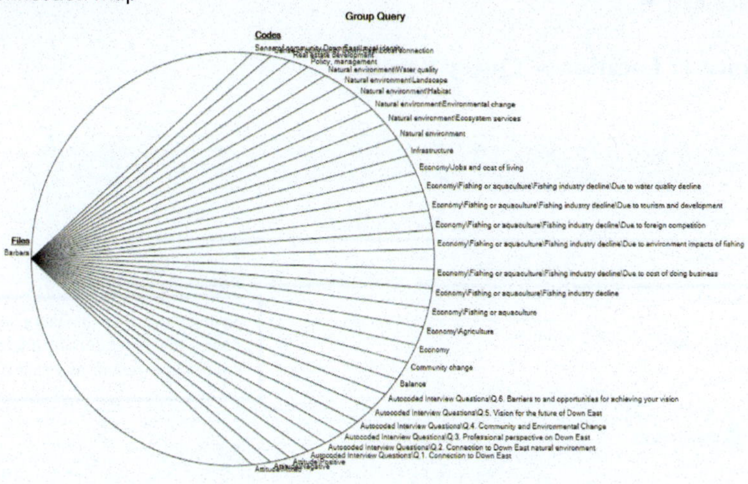

Fig. 12.33 Tabular and graphic output of the group query

Coding With Queries (Save Results)

This paragraph looks at how query results are saved in your project. As we have mentioned before, the results of a query will never be saved as new data, as that would imply data duplication in the project. Never will NVivo duplicate existing data in the project. That means the button *Save Results* in the query window has a different purpose. The button will produce two kinds of outputs: (1) it saves snapshots of query results to the folder *Query Results* (available in *Navigation View* > **Queries** > **Query Results**), and (2) it allows one to immediately apply a code to the results in your codebook (available in *Navigation View* > **Coding** > **Codes**) or create a Case (available in *Navigation View* > **Cases** > **Cases**).

To illustrate this double use of the *Save Results* button, we use a Text Search Query whereby we search for the term *community* in all interviews (we selected the folder *Interviews*), asking to spread the results to the *Broad Context* (see Section "The Text Search Query" for more info on the Text Search Query). After running the query, we found 64 references in 28 files.

When we click on the button *Save Results,* we see the window in Fig. 12.34. The first choice in the window determines whether or not you want to create a new Results item or whether you want to add the results to an existing item. The third option asks you to give the new item a name. When you select to save the results as an existing item, NVivo asks you to select the item to which the results need to be added. For this paragraph, we are interested in the first choice (new item) and, more specifically, the option Location. As shown in Fig. 12.34, the first choice, *Query Results,* is also the standard choice for NVivo when you want to save query results. Therefore, we start to explain this choice first.

Choice 1: Location = Query Results

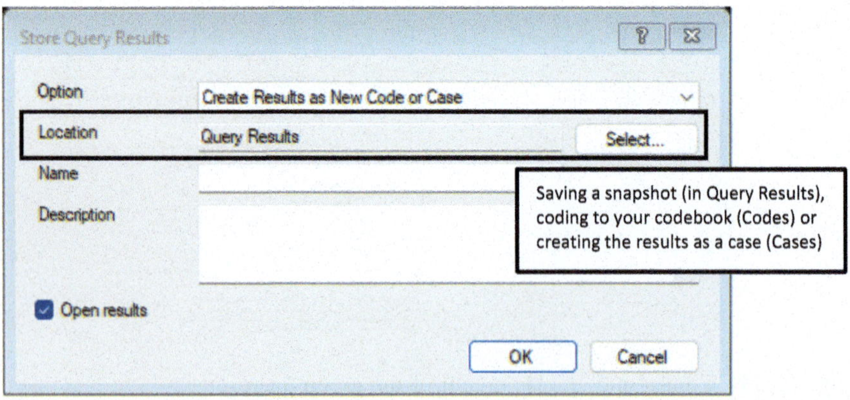

Fig. 12.34 Saving results from a query

When you save your results to the folder Query Results (*Navigation View* > **Queries** > **Query Results**). This folder contains snapshots of NVivo's results when you execute a specific query. A snapshot means that the project item you create in this folder saves the same results of your query when you have executed it and stores it in Query Results. Put differently, it copies the results from your active query and stores them precisely like they are right now.

Some users are not always aware of the implications this has. First, the results store the query definition at the moment of saving but do *not* allow to change this definition any longer. If you want to change the query, you first should have saved the query definition in your project (see Section "Making Your First (Coding) Query"). Consequently, if you do not save the query definition but only save the query results, you can get no updated results of this query in a later stage of your project. What you saved in Query results remains the result of how your project looks *today*. There is no way to convert the query definition of an item in Query Results back into an editable query. You can only open the query definition and look at how this query was created when you saved the snapshot. All options to change that query will be unavailable (greyed out).

Even though the items in Query Results are snapshots, you can convert them into real Codes or Cases by **RMC** (above the item) > **Create As** > **Code** (or **Case**). The same caveat applies here: what is saved to the code or case is what was available as results *when you* created these Query Results. The snapshot does not capture any updates in the project and will also not be converted to the code or case.

Choice 2: Location = Codes or Cases

When you click on the Select button of the option *Location*, you can save your results as a Code in your Codebook (in *Navigation View* > **Coding** > **Codes**) or as a new Case (in *Navigation View* > **Cases** > **Cases**). By navigating the Select Location window, you can save the results to a new child code in your codebook or a new child case. All references found with the query will then be coded with this new code or case.

We repeat that this coding work is only done *once* at the moment of execution of the query. An automatic update of the coding work is foreseen if you re-run the (saved) query definition again later.

Reference

Neuendorf, K. A. (2017). *The content analysis guidebook* (2nd ed.). Sage. https://doi.org/10.4135/9781071802878

CHAPTER 13

Visualising Data with Maps

Key messages in this chapter

- Maps help you visualise your project's (coded) material.
- NVivo provides three different maps with different ways of visualising data.
- The Mind Map is used to brainstorm and organise your ideas.
- The Concept Map helps you to develop theoretical concepts.
- The Project Map visualises parts of the data in your project.

Introduction

During the analysis, a qualitative researcher will construct a theoretical model in a stepwise manner. Concepts take shape and are elaborated, and relationships between concepts are established and documented. This (slow) creation process is often done on paper or in a word processor. Step by step, the researcher develops their ideas until they are ready to be written out in a publication.

NVivo offers the researcher an electronic tool for this stepwise work in the NVivo project. The advantage is that the step-by-step work is preserved within the project, and the researcher immediately has all the project items at hand when building theoretical models. The only drawback is that the modelling tool in NVivo does not always work as flexibly as expected and is rather limited compared to current-day drawing programs available in, e.g. Office365.

D. Mortelmans, *Doing Qualitative Data Analysis with NVivo*, Springer Texts in Social Sciences, https://doi.org/10.1007/978-3-031-66014-6_13

NVivo calls these visualisations Maps and stores them in *Navigation View* > **Maps**. The program includes three visualisations: the mind map, the concept map and the project map. Each visualisation has different graphical tools to build a visual representation of your work. This chapter gives an overview of how to create each of these three types of maps.

The Mind Map

A mind map is a classic brainstorming tool that aims to visualise the development of ideas. The tree metaphor is sometimes used to explain how mind maps are created. The basic idea is to have one central concept in the middle of the mind map. That concept is explored in the brainstorm and visualised in the map. From that central concept some major themes related to that central concept are drawn with "branches" (*Sibling ideas* in NVivo). All major elements connected to the central idea get their branch. Less important elements in the brainstorming are represented as "leaves" to these major branches (*Child ideas* in NVivo) and show up in a smaller graphical representation than the central idea and the major themes. If ideas can not (immediately) be connected to the central elements, they are introduced as free-floating ideas in the mind map (*Free Floating Ideas* in NVivo).

In what follows, we will show a step-by-step construction of a mind map in NVivo (Fig. 13.1). We re-create the Mind Map "early thoughts on coding" from the sample project to do so. We start with **Explore > Maps > Mind Map**.

The first important element (1) is to make your map editable. You see a check box *Edit* in the left upper corner of the map window. To create or change your map, you need to check this box. Otherwise, the map will be protected. When creating a new Mind Map, NVivo automatically puts the central idea on the map as a (blue) circle (2). In the middle of the circle, you can type the name of that central concept. In the example, the central concept is "Change Down East". Keep the blue circle selected and choose *Sibling Idea* in the menu **Mind Map > Sibling Idea**. A rectangular shape with a different colour is added to the map (3). Again, you can label this icon to give meaning to the idea on the map (in the example, community). You repeat this process for all other main ideas while keeping the Central Concept selected (4). We now have one Central Idea (Change Down East) with three connected main ideas (community, economy, and natural environment). Important to note is that each "branch" (Sibling Idea) gets a different colour (after three branches the same colours seem to return).

The next step is adding the less important elements as leaves of the main ideas. To do so, you select the Sibling Idea (5) and click in the menu **Mind Map > Child Idea** (6). You can add as many child ideas as you want (7). For child ideas, the colour of the rectangular is similar to that of the main idea. In that way, NVivo visually shows that these ideas are not new ideas but smaller ones that are under this main idea.

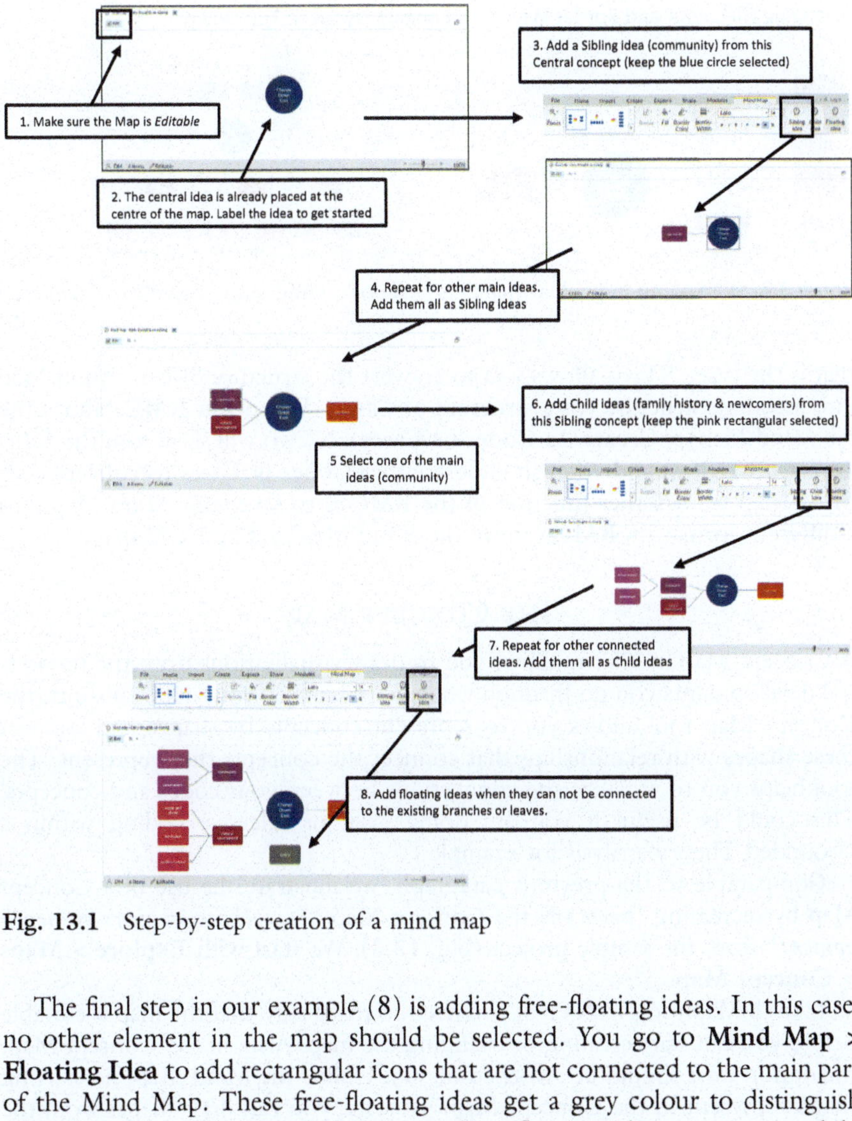

Fig. 13.1 Step-by-step creation of a mind map

The final step in our example (8) is adding free-floating ideas. In this case, no other element in the map should be selected. You go to **Mind Map > Floating Idea** to add rectangular icons that are not connected to the main part of the Mind Map. These free-floating ideas get a grey colour to distinguish them from the other parts of the map. Free-floating ideas cannot get child ideas.

Two final elements to be mentioned involve the layout of the mind map. In the menu **Mind Map**, you find icons to change the fill colour, the border colour and the border width of each element in the Mind Map. Also, three icons determine the overall layout of the map. Standard, the layout is done horizontally, called Mind Map. But you can change this to Top-down or Left–Right (see) (Fig. 13.2).

As Mind Maps are often used early on in a project, researchers tend to use the Mind Map tool to brainstorm about the structure of their codebook. If

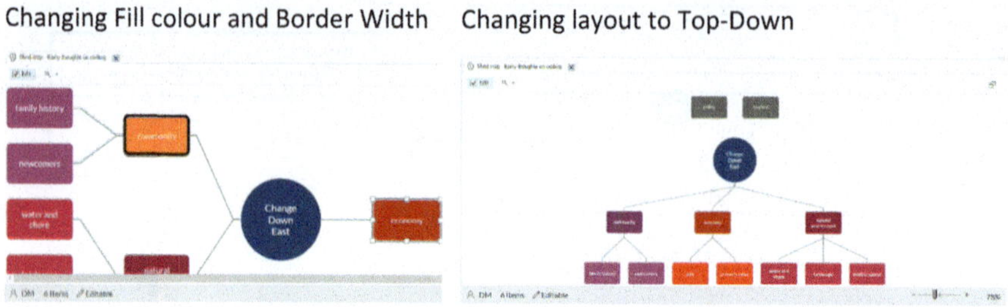

Fig. 13.2 Changing individual elements (left) or a mind map's overall layout (right)

that is the case, NVivo allows you to convert the structure of your Mind Map into a coding structure (or a structure of Cases). When you **RMC** (above the tab Mind Map) > **Create As Codes or Cases**). NVivo will show you the *Select Location* window, and you can choose either Codes or Cases to convert your Mind Map to. Another function of the RMC is to save your Mind Map into a different format by **RMC** (above the Mind Map) > **Export Map**).

The Concept Map

A Concept Map in NVivo allows you to draw visualisations from the theoretical developments you go through while working with your data. In short, the Concept Map tool allows you to represent concepts by shapes and connect these shapes with relationships that connect the concepts they represent. The tool helps you to visualize the connections between your codes and concepts. This could be useful if you are doing axial or selective coding within a Grounded Theory analysis for example.

Comparable to the previous paragraph, we illustrate the use of a Concept Map by recreating (parts of) the Concept Map "*complexity of views on development*" from the sample project (Fig. 13.3). We start with **Explore > Maps > Concept Map**.

Like all Map tools in NVivo, you need to make sure that the map is editable before you can start creating or changing the map. Also, in the Concept Map, the option *Edit* should be switched on (1). Unlike the Mind Map, no starting shape is present in the Concept Map. To start, you can drag shapes from the left panel into the drawing area (2). Next to shapes, you can also add Project Items to your Concept Map. These shapes are automatically determined by the icons they have in the program. When you add codes, NVivo will use the black circle as the shape of these items in the Map. To add project items, click in the menu **Concept Map > Add Project Items** (3). In the *Select Project Items* window, you select the items you want visualised in the Concept Map. In the example, we have chosen two codes from the codebook (4). Each project item gets its own icon in the Map (the same icon as is used in the main program).

The next step in creating the Concept Map is linking two elements with connectors. You have three connections: one-way, symmetrical, and associative

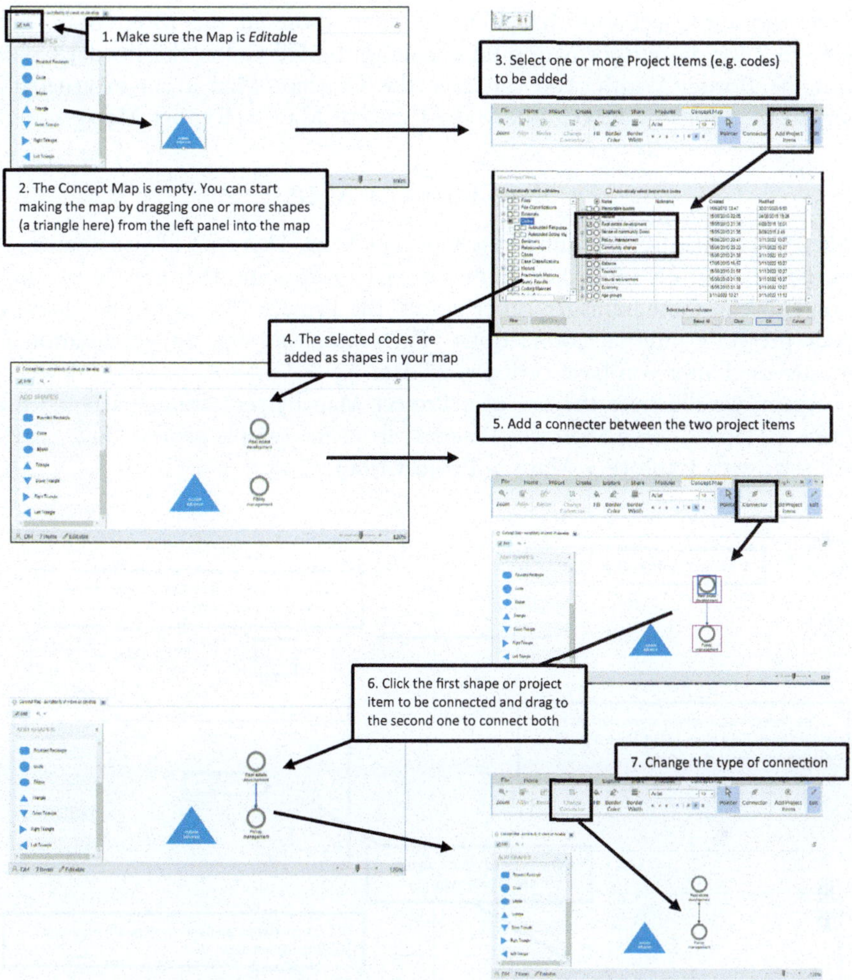

Fig. 13.3 Step-by-step creation of a concept map

(see Section "Linking Data with Relationships" in Chap. 11). To link shapes, click on **Concept Map > Connector** (5). Next, you select the first item to be linked and immediately drag towards the second shape (6). When you release the mouse above the second item, both are connected. NVivo will draw a one-way-shaped connecter. To change this connection, select the connection arrow and click on **Mind Map > Change Connector**. In the drop-down menu, you can select the connector type you want. To stop adding connectors and add or change the shapes again, click on **Concept Map > Pointer** to shift the focus from connecting to adding and changing shapes.

The concept map's colour or layout of project items can not be changed. They remain the standard icons the program foresees. The shapes you select from the left panel can change similarly to those in the Mind Map. See Fig. 13.2 for more information on changing the Fill Color, the Border Color and the Border Width. You can save the Concept Map in an external file format (e.g. jpg) by **RMC** (above the Concept Map) > **Export Map**).

The Project Map

The Project map is a variant of the Concept Map. This type of visualisation is focused exclusively on project items of your project and the internal connection between these items. In other words, the Project Map helps you visualise your project's internal connections. Therefore, this type of visualisation is mostly used after you have coded your data.

Again, we illustrate the use of a Project Map by recreating (parts of) the Project Map "*Coding Structure Progress*" from the sample project (Fig. 13.4). We start with **Explore > Maps > Project Map**.

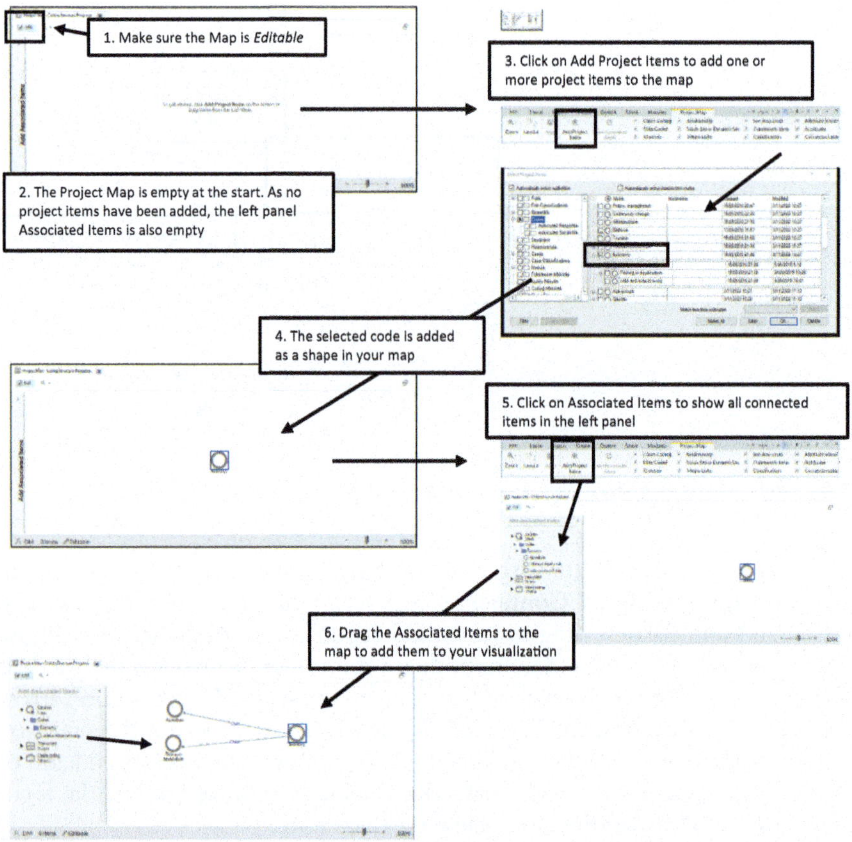

Fig. 13.4 Step-by-step creation of a project map

As with all maps, you first select the *Edit* option to make your map editable (if not yet selected) (1). Like the Concept Map, the Project Map starts with an empty canvas (2). As you need to start with some Project Items to start the map, you click on **Project Map > Add Project Items**. In the *Select Project Items* window, you select one or more project items from your project that you want to start with (3). These can be Files, Codes, cases, etcetera. There is no preferred type of project item that you should start with. NVivo will place the selected items on your map (4). Next, you need to expand the map by adding items connected to the ones you already have available. To do so, select one of the shapes in your map and click on **Project Map > Show Associated Items** (5). In the left panel, all items connected to the one you had selected are shown now. You can now start dragging associated items to the map from that left panel. These will be added with a connector that identifies the relationship between the two items. E.g., if you start with a code and add child codes to the map, the connectors will say "Child" to indicate that the new codes are child codes from the original code. If you add a file coded with that code, the connector will say "Codes".

Concerning the layout, we refer to the options discussed in Fig. 13.2. The shapes can not be changed as project item shapes are always represented with the standard icons of the program. You cannot change these. But the menu Project Map > Layout can change the overall layout. Four automatic layouts are available: Layered Directed Graph, Hierarchical, Circular, and Directed. It depends on your taste which of these options fits your visualisation the best. The map can be exported to an external graphical file by **RMC** (above the Project Map) > **Export Map**).

CHAPTER 14

Reporting and Exporting Data

Key messages in this chapter

- Reports are pre-structured and standardised overviews of the content of your report.
- You can develop your custom-made reports.
- Text Reports (Extracts) allow you to export (parts of) your data to external programs.
- The REFI-QDA format allows the exchange of project elements between QDA programs.

Introduction

You may need to get an overview of your project and its content at a certain point in your analysis. One way of obtaining specific views on your data is by using the Matrix Coding Query (see Section "The Matrix Coding Query" in Chap. 12). In this chapter, we show a second way of summarising data in your project: reports. Next to these summaries, we also discuss ways to export your data both at the level of the project item and on a larger level with extracts. We will for example show how you can export your codebook to a PDF or to another QDA program for a colleague (see Sections "Formatted Reports" and "Exporting Project Material with Text Reports (Extracts)").

D. Mortelmans, *Doing Qualitative Data Analysis with NVivo*, Springer Texts in Social Sciences, https://doi.org/10.1007/978-3-031-66014-6_14

Formatted Reports

In *Navigation View* > **Reports** > **Formatted Reports**, you can find eight reports NVivo offers users. In Table 14.1, we give an overview of the content of each of these reports. It suffices to double-click on the name of one of these reports to create and see the report. If you want to save the report outside NVivo, simply **RMC** (above the report) > **Export Report Results**). Reports can be saved as Word documents (docx, rtf), PDFs, Excel files and Web Pages (.html).

Next to these built-in reports, users can make their own reports. Two tools from **Share** > **New Formatted Report** are *Wizard* and *Designer*. The Wizard leads you step by step through making your report. The Designer is for advanced users. It presents an empty report and lets you configure the report all by yourself. We cannot give a comprehensive overview of all possibilities for creating reports using the Wizard or Designer, but we will summarize the main elements of the creation process.

Table 14.1 Overview of built-in reports in NVivo

Report name	*Description (as mentioned in formatted reports properties*[a]*)*
Case classification summary report	This report shows how many cases are within each classification, including a breakdown by attribute value for each attribute
Code summary report	This report provides a list of Codes, including summary information about the files or externals coded at each
Coding structure report	This report lists codes by type. It includes nicknames and user-assigned colours
Coding summary by code report	This report summarises coding, grouped by code, then file or external
Coding summary by file report	This report summarises coding, grouped by file or external, then code
File classification summary report	This report shows how many Files or Externals are within each classification, including a breakdown by attribute value for each attribute in the classification.
File summary report	This report summarises the coding for files or externals by code
Project summary report	This report lists folders in a project and the items they contain

[a] The explanations in this column are literal citations of the field *Description* and are not written by the author of this book.

The Wizard

Start the wizard with **Share > New Formatted Report > New Formatted report via Wizard**. In total, NVivo asks for information in eight steps. In Fig. 14.1, we present the eight steps without discussing them more deeply.

The Designer

The designer gives complete freedom to the user to create their custom-made reports. You start the Designer by **Share > New Formatted Report > New Formatted report via Designer**. In the *New Report* window, you must provide basic information about the report to start the design process (Fig. 14.2).

When NVivo has the basic information on the report, a tab is opened with the first page and a right-hand panel with all fields you can add to the report (Fig. 14.3).

By dragging fields to the page layout, you create your report step by step. In the menu **Designer**, multiple tools are offered that were also present in the wizard: sorting data, grouping data, and filtering data. You get a look at the result by **Designer > Run Formatted Report**. A new tab is opened in that case, showing the current report with all the data filled in. You can always return to the designer tab to edit and re-run the report to see the result.

Exporting Project Material with Text Reports (Extracts)

In the previous paragraph, we discussed formatted reports. These are reports with data from our project in a formatted layout. In this paragraph, we discuss so-called Text Reports (as they were called in previous versions: extracts). The name Text Report is confusing as there are no real reports generated in the sense as we saw with the formatted reports. A Text Report is a tabular extract of data intended to export the data from your project to external databases (in the form of an Excel file). Table 15.2 gives an overview of what the Text Reports (in Excel) look like after they have been exported. For the column names, we highlight the most important fields of interest in bold. The analytical unit is important as it determines the type of data management you can perform on the dataset afterwards to analyse the data with statistical programs.

The built-in Text Reports give any data export in any direction you want. However, NVivo chooses to export a high number of fields (see the third column in Table 14.2). If you want control over what is exported, you control your Text Reports with a Text Report Wizard starting at **Share > New Text Report**. This wizard (see Fig. 14.4) consists of only four steps, which are similar to the ones we discussed in Fig. 14.1. In the first step, you again need to choose the analytical unit of your export (see also column 2 in Table 14.2).

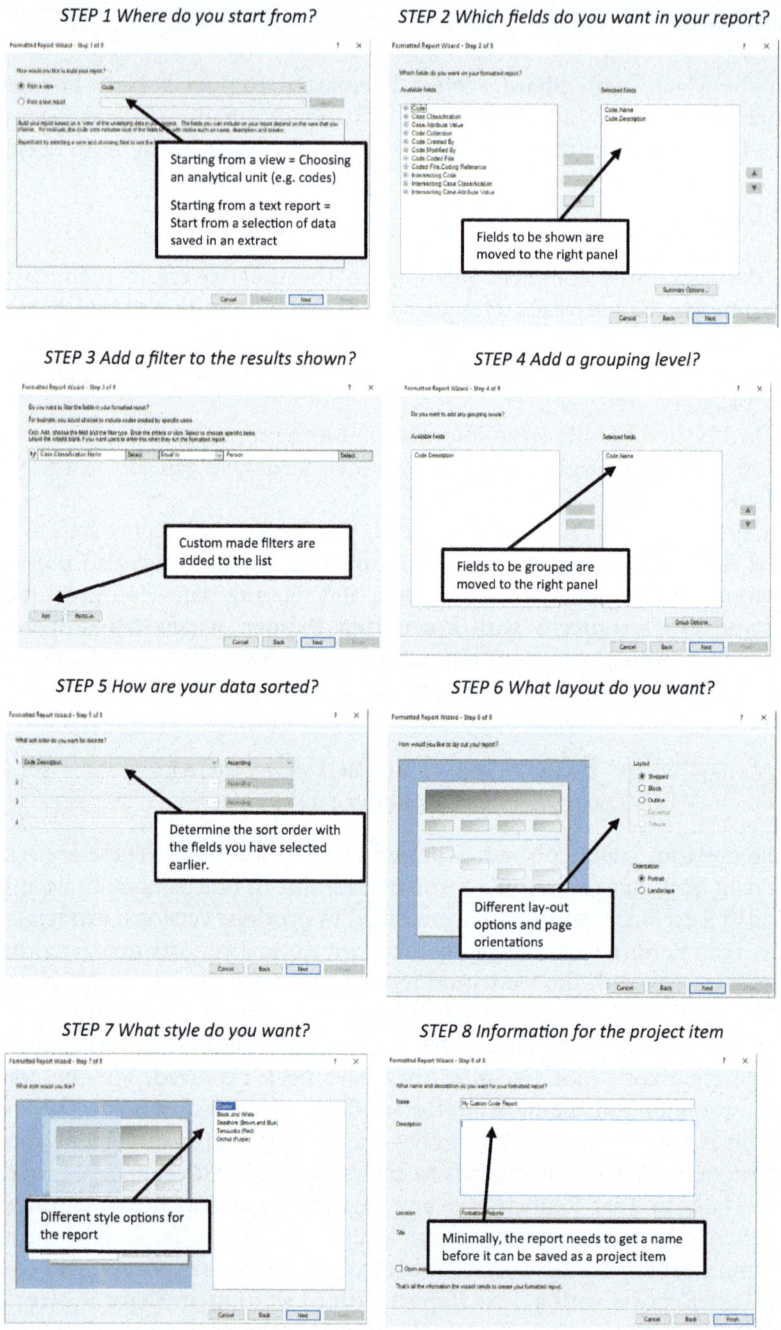

Fig. 14.1 Using the formatted report wizard

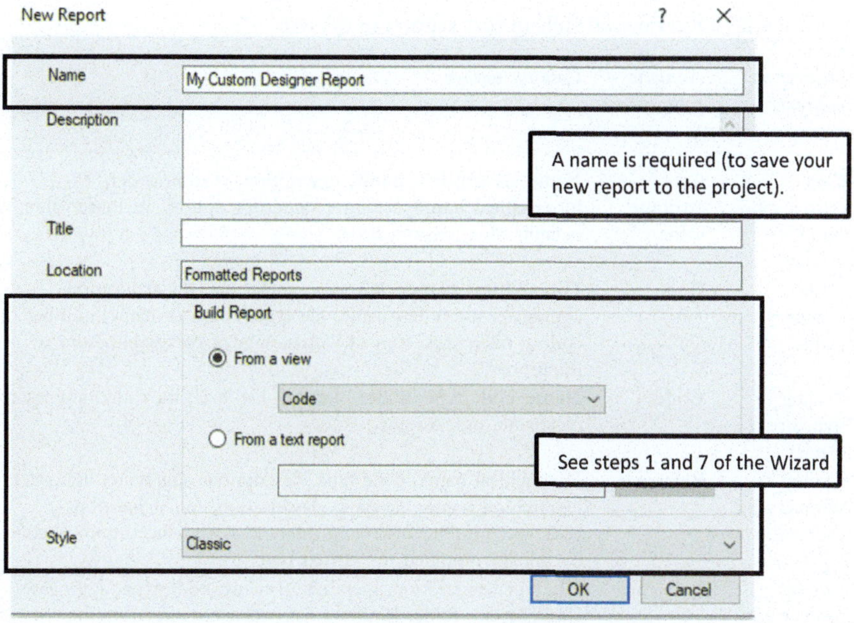

Fig. 14.2 Using the Designer for building customised reports

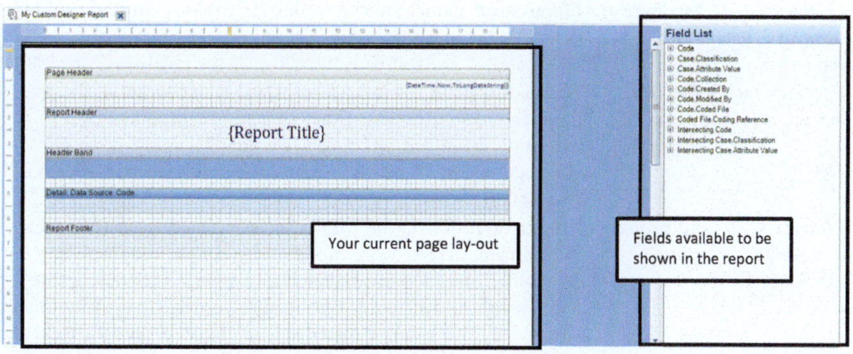

Fig. 14.3 The Designer tab for building customised reports

When you click Finish in Step 4, NVivo will save the Text Report as a project item to your project, immediately run the report, and export your data. We get an Excel file shown in Fig. 14.5. The Excel file is similar to the built-in Report *Coding Structure Extract* (see Table 14.2 Overview of built-in Text Reports in NVivo), but the file now has only three columns (*Name, Description, and Code Type)* instead of eight.

Table 14.2 Overview of built-in text reports in NVivo

Text report (extract) name	*Analytic unit (rows)*	*Column names*
Case classification summary extract	Case × attribute value	Name, description, **name**, description, custom order, **hierarchical name**, attribute value description, attribute value, is default value, custom order
Code summary extract	Code	**Hierarchical name**, code type, nickname, folder location, aggregate, list order, name, file type, count of name, number of coding references, sum of words, sum of paragraphs, sum of duration
Coding structure extract	Code	**Name**, code type, folder location, list level, list order, aggregate, nickname, user assigned color
Coding summary by code extract	Reference	Hierarchical name, code type, description, aggregate, list order, hierarchical name, file type, classification, folder location, coverage, number of coding references, reference number, coded by initials, modified on, **coded text**
Coding summary by file extract	Reference	File type, description, hierarchical name, code type, aggregate, classification, folder location, list order, coverage, number of coding references, reference number, coded by initials, modified on, **coded text**
File classification summary report	File × Attribute Value	Name, description, **name**, description, custom order, file type, **hierarchical name**, attribute value description, attribute value, is default value, custom order
File summary extract	File (or memo or external)	**Name**, hierarchical name, description, folder location, file size, file type, created by username, created on, Modified by username, Modified on, words, paragraphs, count of name, coverage, count of words, count of duration, Count of region Proportion
Project summary extract	All project items	Project title, description, file location, created by initials, created by username, created on, modified by initials, modified by username, Modified on, hierarchical name, name, description, root system folder, folder type, project order, Created by initials, Created by username, Created on, modified by initials, modified by username, modified on, Hierarchical name, **name**, **item type**, list level, list order, created by initials, created by username, created on, Modified by initials, Modified by username, Modified on

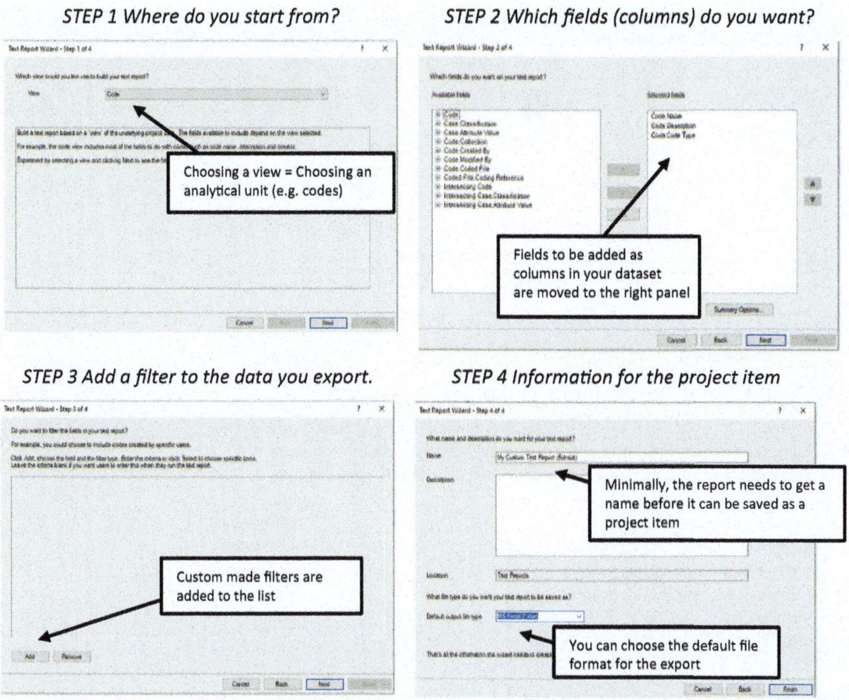

Fig. 14.4 Using the Text Report wizard

Exporting Project

Copying an Entire Project

NVivo has no automatic backup function. Therefore, it is important as a user to make back-ups of your project file (.nvpx) yourself. Backup copies can be created when the program is closed in your Explorer. The Copy Project option can also be done from within the program. This option is found in two places in the program. First, you can make copies by **File > Copy Project**. Second, the export is also found in **Share > Copy Project**. The *Copy Project* window is shown (Fig. 14.6). You can copy the current project to copies in the Windows, MAC, and Server formats. So, backup copies can be made, and you can also export the project to a different NVivo project format to exchange it with colleagues working on another platform. The MAC version however, can only save as a MAC file and cannot convert to the Windows or server formats.

Export Parts of a Project

In **Share > Export Project**, you also have the opportunity to export the entire project. But rather than copying the whole project, you can select particular project items you want to export. This can be important if you want

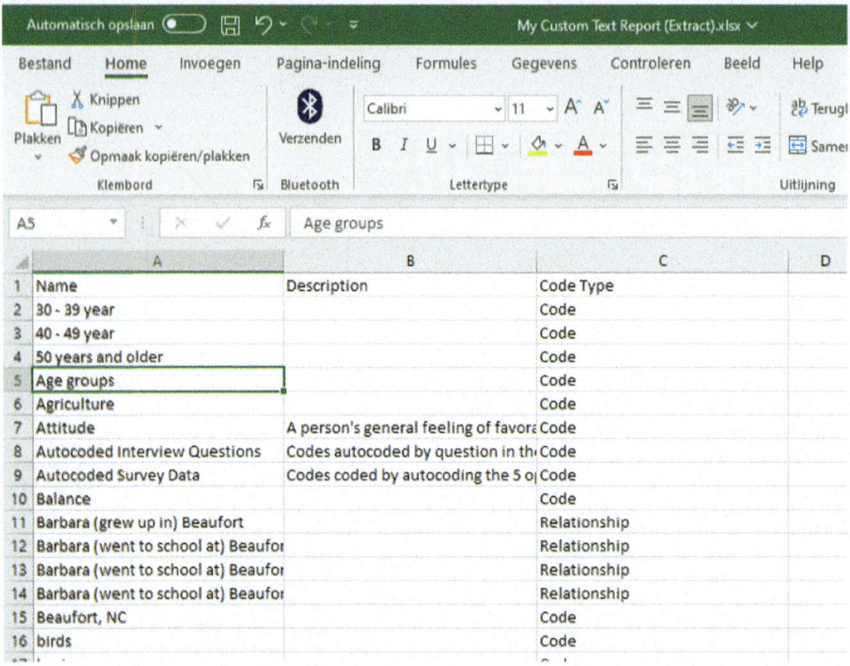

Fig. 14.5 Export of data to an Excel file

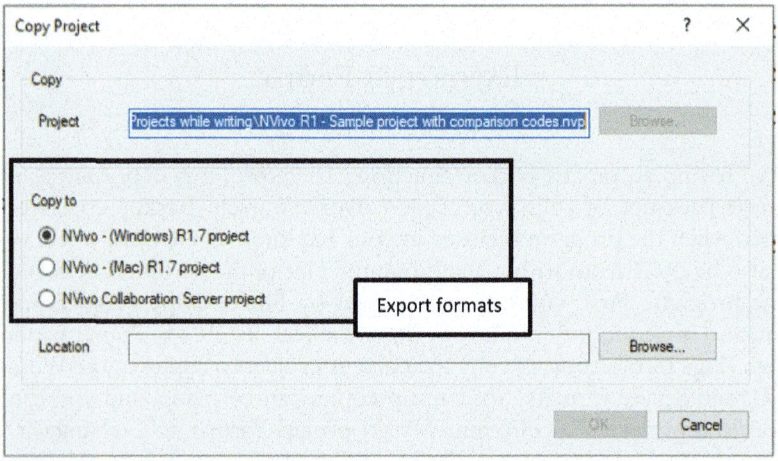

Fig. 14.6 The Copy Project window

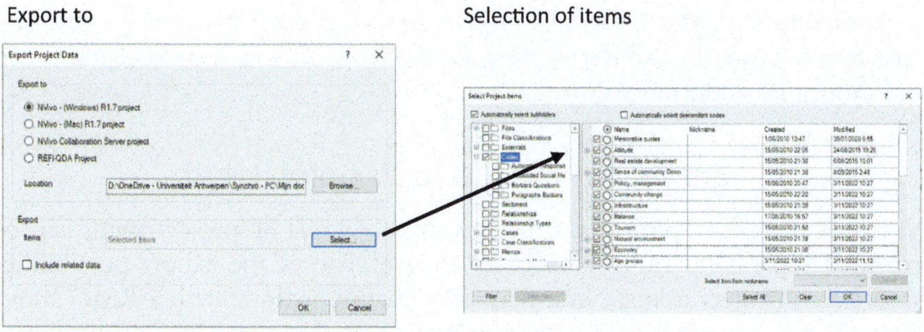

Fig. 14.7 Making selections (the codebook) while exporting a project

to pass on your codebook (and only your codebook) to another project or give it in NVivo format to a colleague. In that case, you can export your project and select the coding structure without the files and the references (see Fig. 14.7). Your colleague who receives this project can then import the project (**Import > Project**) and receive your codebook without any other elements from your project.

Export to the REFI-QDA Format

An important option to mention in the **Share > Export project** window is the REFI-QDA Project format. REFI stands for *Rotterdam Exchange Format Initiative* (see qdasoftware.org for more information). The REFI-QDA standard is a software-independent file format that either contains complete QDA projects in a software-independent format (.qdpx and .qde) or contains the codebook of a project (.qdc).

In the first phase of this project, the REFI team concentrated on exchanging codebooks across software programs. The export of entire projects somewhat surpasses the feature, but if you want to export your codebook only to the .qdc format, do so. **Share > Export > Export Codebook**. NVivo will suggest exporting to docx-format, but when you click the *Browse* button, you can also export to the .qdc-format. If you want to import a codebook in the .qdc format, go to **Import > Codebook** and import the codebook.

At present, NVivo can also export (almost) entire projects to the REFI-project standard (.qdpx) in the **Share > Export project**. Again, the user can select particular items that need to be exported to this format. When you receive a project made in another software program and want to import it, use **File > Open** and let NVivo convert it to its native format. Note that you cannot import the REFI-QDA files with **Import > Project**. This type of import is reserved for NVivo project files only.

As mentioned before, the REFI-QDA standard is still in development. You can always check the NVivo website for updates on which project items can and cannot be exported to this format.

Exporting Specific Project Items

Not only the entire project can be exported. Almost all project items can be exported separately. Table 15.3 gives an overview of the export possibilities of different project items. Note that all exporting is done with a RMC above the item you want to export. The export option in the RMC menu always starts with Export. We will not repeat these options in Table 14.3 unless there is doubt about the option you must choose (sometimes, two or more export options are available).

Table 14.3 Export possibilities of individual project items

Project item	*RMC target*	*Export-format(s)*	*Remark*
Folder	Folder > export list	txt, docx, rtf, pdf, xlsx	A list of all items in a folder is exported
File–document	File > document-file	txt, docx, rtf, pdf, html	
File–PDF	File > pdf-file	pdf, html	Also, options are offered to export to reference managers
File–audio	File > audio-file	HTML, mp3	Also, options are offered to export the transcript to docx
File–video	File > video-file	HTML, mp4	Also, options are offered to export the transcript to docx
File–picture	File > picture-file	HTML, jpg	Also, options are offered to export the log to docx
File–dataset	File > dataset	xlsx, txt (tabs), HTML	NVivo offers to add a unique ID to the dataset and can also export column headers
File classification			Not exportable as such (see file classification sheet)
File classification sheet	File classification sheet tab	xlsx, txt (tabs), sav	The SAV format is an export to the IBM SPSS statistical software program
External	Externals > external	txt, docx, rtf, pdf, html	
Code	Codes > code	txt, docx, rtf, pdf, html, xlsx	NVivo offers three options: entire content, reference view, and summary view
Codebook	Codes	docx, xlsx, qdc	More information on the QDC format is in Section "Export Parts of a Project". When you have a codebook in a folder under Codes, this folder can be exported as a separate codebook
Case	Cases > case	txt, docx, rtf, pdf, html, xlsx	NVivo offers three options: entire content, reference view, and summary view
Case classification			Not exportable as such (see file classification sheet)
Case classification sheet	Case classification sheet tab	xlsx, txt (tabs), sav	The SAV format is an export to the IBM SPSS statistical software program

(continued)

Table 14.3 (continued)

Project item	*RMC target*	*Export-format(s)*	*Remark*
Memo	Memos > memo	txt, docx, rtf, pdf, html	
Framework matrix	Framework matrices	xlsx, txt	
Annotations			Not individually exportable. Only a list of all Annotations can be exported
See-also-links			Not individually exportable. Only a list of all relationships can be exported
Sets			Neither static nor dynamic sets can be exported individually
Query (definition)			Not individually exportable. Only a list of all Query names can be exported
Query (results area)	Above the results area after running a query	Format dependent on the type of query	Run a query and RMC above the results area to see to which format the results can be exported. Textual output can also be selected in the results area and copied with CTRL + C
Map	Maps tab	jpg, bmp, gif, png, tif	Beware that none of these formats are vector formats. All formats are bitmap-based formats
Formatted report	Report tab	txt, docx, rtf, pdf, html, xlsx	Only opened reports can be exported
Text report definition	Text reports > Text report	Nvx	More information on the NVX-format in Section "Exporting Project Material with Text Reports (Extracts)". The NVX format contains the definition of the Text Report and not the results (see next line)
Text report	Export while running the extract	xlsx, txt, XML	

CHAPTER 15

Using NVivo in Focus Group Research

Key messages in this chapter

- Data from Focus Groups are analysed in NVivo with the same tools as data from interview research.
- Unlike interview research, focus group research has two analytical levels: the group level and the participant level.
- To analyse focus groups, you create case classifications at both the group level and the individual level.
- The Text Search query can be helpful in coding answers of participants to cases.
- The Matrix Coding Query helps to make comparisons both at the group level and at the individual level.

Analysing transcripts of focus groups differs from analysing interview material. To a large extent, analysing focus groups amounts to applying the tools already covered earlier in this book. New tools are therefore not discussed in this chapter. However, we will reflect on how to use NVivo when analysing focus group transcripts.[1] After all, NVivo's flexibility means that the richness of focus

[1] In this chapter we approach the analysis of focus groups from the perspective of analysing focus groups transcripts. This is done to show the parallels and differences with the discussion of interview transcripts in earlier chapters. Often focus groups are analysed both from a textual transcript and from video material that allow analysing gestures and

D. Mortelmans, *Doing Qualitative Data Analysis with NVivo*, Springer Texts in Social Sciences, https://doi.org/10.1007/978-3-031-66014-6_15

group discussions can also be fully exploited by efficiently using the existing tools.

Some Background on Analysing Focus Groups

The father of the focus group, Robert K. Merton (1946, 1987, 1990), defined the focus group as a research technique used when a group of respondents has collectively experienced something and is interviewed about this experience. The idea behind this was based on the communication science research he conducted, which had in the back of his mind "seeing a film or an advertising campaign" as the joint stimulus of focus group participants. Today, the core of Merton's early definition remains: focus groups bring together people who share a common characteristic to talk about a particular research topic under the guidance of a moderator.

Unfortunately, the enormous popularity of the focus group technique means that focus groups are used appropriately but just as often inappropriately (Greenbaum, 1993; Ledingham & Bruning, 1999). The most common problem is that focus groups are perceived as a cheap alternative to quantitative research. Instead of a survey of several hundred or thousands of people, people revert to a few focus groups to gather information about a particular population group. This happens not only by companies in a commercial setting but sometimes also by scientific institutions faced with shrinking budgets made available for scientific research. Focus groups can't replace large survey samples, just like it can't be used as a low-cost alternative to qualitative face-to-face interviews. In that case, too, the relatively low cost of focus groups incentivises the misuse of the technique. It is not because one brings some people around the table that you have done a proper face-to-face interview with each of them. Focus groups need to be analysed both on the individual and on the group level, as we will show in this chapter. In our opinion it is wrong to regard a focus group as a collection of individual interviews that happened to be conducted simultaneously.

In focus groups, people influence each other, and there is a development in the group discussion. This is fundamentally different from a face-to-face interview. Therefore, the core of the analysis in focus groups is that the group is the unit of analysis and not the individual. In what follows, we will discuss (four) different strategies to analyse focus group data that all respect the group as a level of analysis.

The first strategy is to count opinions. This involves checking how often (frequency) and widespread they are in the focus groups. The **frequency** of statements is the number of times a particular idea recurs in the discussion. It does not mean "applied by different people". When a particular respondent brings up the same argument repeatedly, the frequency of this argument

non verbal behaviour in the groups. We refer to Chap. 16 for the analysis of multimedia material as a supplement to the text-based analysis techniques introduced in this chapter.

increases. To make a correct estimate of the frequency of opinions, it is crucial to have transcribed texts with the names of the participants (see Section "Transcribing Focus Groups"). Otherwise, the bias may be too large. The **spread** of statements examines how many participants made a certain point. So here, you look at who makes arguments. Checking the spread is also done across groups. When the researcher wants to compare different group profiles, they can check to what extent people put forward the same arguments and what the spread of these ideas is within the different groups. The problem with this is that counting statements can never be sufficient to determine how important an idea is (Morgan, 1995). Suppose five people in one group have the same idea on a topic, but that idea does not surface anywhere in the other groups. Does this idea then count as one idea because it only surfaced in one group, or do you still count that idea as five? The reverse can also happen. Suppose that in five groups, one individual puts forward a particular idea but does not find support anywhere among the other participants. Here, an idea is put forward in five groups, but the idea does not find any support in these groups. Again, it is difficult to claim that there is a spread of the idea in five groups.

Next to counting, attention to the conversational context is crucial. Statements in focus groups can only be understood in the context of the conversation in which they were made. This might not be new, as context is central to qualitative analysis. Nevertheless, the researcher needs to pay extra attention to the context of the conversation when analysing focus groups because they should remain very aware that people's statements could have been influenced during the conversation. Therefore, it is wise to identify broader text references during the open coding phase so that statements are not separated from the context in which they were made. If available, visual material like video recordings of the focus group help in understanding the conversation context. Looking at the way in which ideas are expressed or how non-verbal behaviour in the group illustrates the interaction between participants, enriches your analysis.

When studying context, the researcher can pay attention to several elements. First, there is the **intensity** with which certain group members make statements. Sometimes, the intensity is evident through word usage that becomes emotional or because people start explicitly emphasising ideas: "I think it is very important that…". In addition, intensity is mainly expressed by the tone in which people speak or the pace of speech or non-verbal gestures to underline their claims. Here, the researcher relies heavily on additional data besides the texts (e.g. visual material or moderator reports). To estimate how much weight a particular statement should be given in the analysis, the researcher can also look at the **specificity** of what is being told (Krueger & Casey, 2014). Are respondents' answers based on experiences, or do they give only vague and general views? The latter are given less weight in the analysis than personal statements—something the researcher can pay extra attention to the use of the first person singular. When a respondent tells a story in the

"I" form, it represents more weight in the analysis than when respondents talk about "everybody thinks like this" or "one does this or that".

Third, you need to look at the **course of the entire conversation**. Here, the context is not focused on a single question, but the researcher looks at the conversation as a whole. After all, influence and change of views need not occur abruptly and immediately. In the analysis, the researcher has to be attentive to changes in viewpoints across the participants. First, he must observe these changes (again, visual clues can help in this case). This requires thoroughly studying a respondent's statements across the conversation (Carey & Smith, 1994). The second step is then to ascertain whether the respondent themself agrees with the change in opinion. A dominant participant can convince some participants seemingly. In further conversation, however, you might see that these participants have reverted to their old views. At that point, you don't weigh the temporary reversal too heavily. The third and final step in the analysis of changing opinions is the search for the cause of the change. This is interesting analytical material. To know what triggers change in people is to gain a deeper understanding of what persuasive and non-convincing arguments are. Possible causes include the argumentation style someone has, new evidence to support an idea, a personal story that presents views in a different light, and so on.

Finally, when analysing focus groups, the role of the **moderator** (Bertrand et al., 1992) is also important to integrate in your analysis. Every moderator has their view on the course of the conversation that is useful in the analysis. Therefore, the moderator is sometimes asked to prepare a methodological report after the conversation, summarising what they think are the most important elements that emerged in the group. As a researcher, you do not take this view as "truth" but consider such a report as additional data that can be analysed. It is a directional report that can provide certain elements (from the moderator's perspective, of course) for analysis and draw the researcher's attention to certain elements. Not infrequently, this involves things that the moderator has seen first-hand, which are difficult to accommodate in the transcripts (e.g. non-verbal behaviour of respondents).

In sum, the analysis in focus groups runs largely parallel to that of other types of qualitative data. That is, the researcher can perfectly use the different tools of NVivo for focus groups. The idea of breaking data down into particles (coding) and rebuilding afterwards (querying) applies perfectly to focus group data. The big difference, however, lies in the attention the researcher will alternate between two analytical levels: the individual and the group level (Parker & Tritter, 2006) and the attention to the internal dynamics of the conversation. The conversation will be viewed from a distance, as it were, as a social fact and analysed as such.

Transcribing Focus Groups

In Section "Transcribing Media Files", we explained how you can transcribe media files in NVivo if you did not transcribe them with your own Word processor. Focus groups are no different from interviews concerning the tools you can use for transcribing. It can be done inside and outside NVivo. One difficulty in transcribing focus groups is that there is no one-to-one conversation. In an interview, both interviewer and interviewee alternate in taking the word. In a focus group, it often happens that different participants are talking at the same time or that they interrupt other participants while they are speaking. That gives a rather "jumpy" transcript with many changes in speakers across the transcript. Also, a group conversation makes it more difficult to position your recording device optimally. It could be that people at one side of the table are harder to understand on the recording because they are located further from the device, especially when they talk quietly.

Besides these technical challenges, you also need to consider the multiple-participant situation in the actual transcript. Therefore, we propose the following transcription rules to facilitate analysing the transcripts in NVivo later on somewhat easier:

1. *Questions and reactions of the moderator are put in a specific paragraph style or bold.*
2. *Use the following scheme for participants:*

A question and answer of the moderator/participant starts with # and ends with a double point

#Moderator:
#NAME: (Using the First Name of the Participant).

When you have many participants with the same name, you make these names unique in the transcript

#John1: #John2:

Each participant's name should be unique throughout the research (also across transcripts of multiple groups).

3. *Names from participants are also put in a specific paragraph style or in bold.*
4. *When participants use each other's (first) names in an answer, you use the name and omit the # before their name.*

The background to these rules is having the participant's name identified uniquely. There are two ways to code answers of participants in focus groups:

with the Text Search Query (see Section "Using Queries in Focus Groups") and with the Autocoding (Speaker name, see Section "Semi-auto coding Based on Speaker Names (in Focus Groups)" in Chap. 19) tool. For these tools to work conveniently, participant's names should be used uniquely and consistently throughout your transcripts. Using only the names of the participants is not sufficient, as they can refer to each other in their answers. Therefore, we advise adding a special symbol like a hashtag (#), an at symbol (@) or a dollar sign ($) before the name of the participant. As such, the participant's name is uniquely different from the answers, where participants can also use each other's names when answering other participants, for example.

When using the Text Search Query, participant's names are put in the paragraph where they answer:

#JOHN: I agree with what is said, but I want to add …

For the Autocode function, the previous works fine, but also a transcript whereby the participant's name is put above the answer:

#JOHN:
I do agree with what is said, but I want to add …

It is also important that you explicitly add the #MODERATOR to all questions and reactions of the moderator of the focus group. Especially when using the Autocode tool, the moderator must be identified as a separate participant for the tool to work. When using the Text Search Query, this is less of an issue.

Comparing Interviews and Focus Groups as Analytical Data Sources

Crucial to the analysis of focus groups is the double analytical level at which you will analyse the data: the level of the individual participant and the focus group. In this paragraph, we show you how these two levels of analysis are found in an NVivo project and how they compare to an interview.

When we look at the *Navigation View* > **Files,** there is not much difference when looking at the transcripts of interviews or those of focus groups (see Fig. 15.1): transcripts are files, and the files are listed in the *List View*. Only the name of the files makes the difference between an interview transcript and that of a focus group.

When we open the transcripts, the difference becomes clear (Fig. 15.2):

In the interview, only two actors operate: the interviewer and the interviewee. Inside the transcript of the focus group, we also have one moderator, but there, of course, multiple participants. This makes a great difference in working with the files (transcripts) of focus groups. For the interview, the analytical level is the interviewee. That means that the file level coincides with

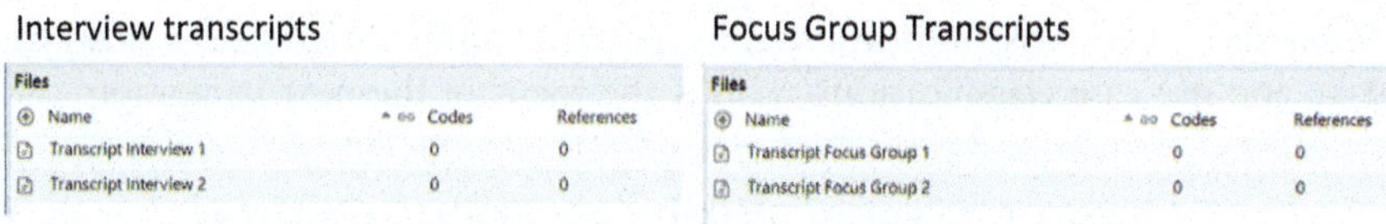

Fig. 15.1 Analytical levels in focus groups 1—the files-level

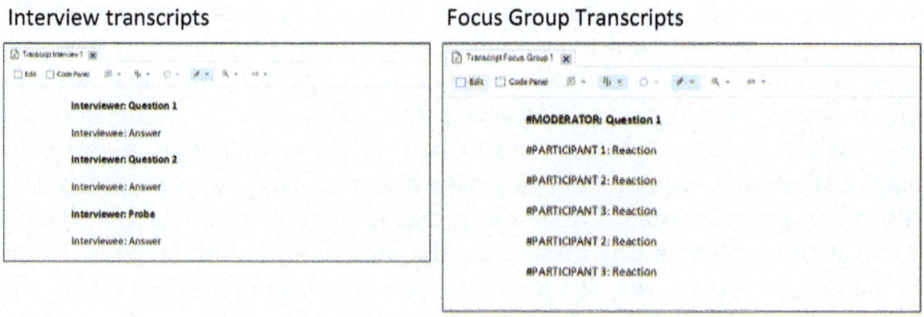

Fig. 15.2 Analytical levels in focus groups 2—inside the files-level

the analytical level. In focus groups, we have the analytical level of the group, which coincides with the file level. But we also have the analytical level of the participant, which does *not* coincide with the file level. The participant level is located *inside* the file level. We show these differences graphically in Fig. 15.3.

An easy example to make the difference clear is *gender*. In an interview, the whole conversation was made with a woman, for example. That means that gender = woman is a characteristic situated at the interview level, or put differently, at the file level. For a focus group, this is not the case if we have mixed-gender groups. Participants 1 and 3 may be women, and participant 2 may be a man. So, gender = woman can not be attributed to the group- or file level. It can only be attributed to an intra-file level of the answers of participants 1 and 3. If you want to make comparisons across genders, for interviews, you will compare files. For focus groups, you will compare answers

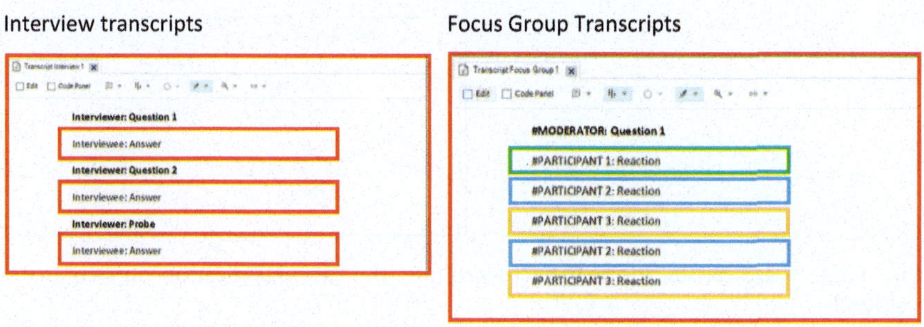

Fig. 15.3 Analytical levels in focus groups 3—analytical levels (coloured)

within files. The most important consequence of this difference is how you make and use case classifications. This is the focus of the next paragraph.

Using Classifications in Focus Groups

In Chap. 9, we introduced file and case classifications as tools that help researchers make comparisons in their analysis more easily. We used examples from interview research in that chapter to explain how classifications work. We did not mention analytical levels in the explanation, as for interview research, these are not crucial to consider. For focus groups, we know that we have *two* analytical levels (group and participant), so we need to create both our case classification and our cases on these two levels. In Fig. 15.4, we show both case classifications for focus group research. First (upper right panel), you make a case classification at the group level. This classification contains stable attributes within one group, like the moderator and the time and place of the group. Second, we also prepare a case classification at the participant's level. Here, we add attributes like gender and age.

Case Classifications are added to cases in our project. As we now have two analytical levels, we need to create cases on these two levels as well (Fig. 15.5). Again, in interview research, one interviewee gets one twin case. In focus group research, we need to create cases for the group level (upper right panel), and we need cases for each participant (lower right panel). To make our life easier, we created two folders in *Navigation View* > **Cases** > **Cases** so that the cases at the group level are stored in a separate folder as the cases of the

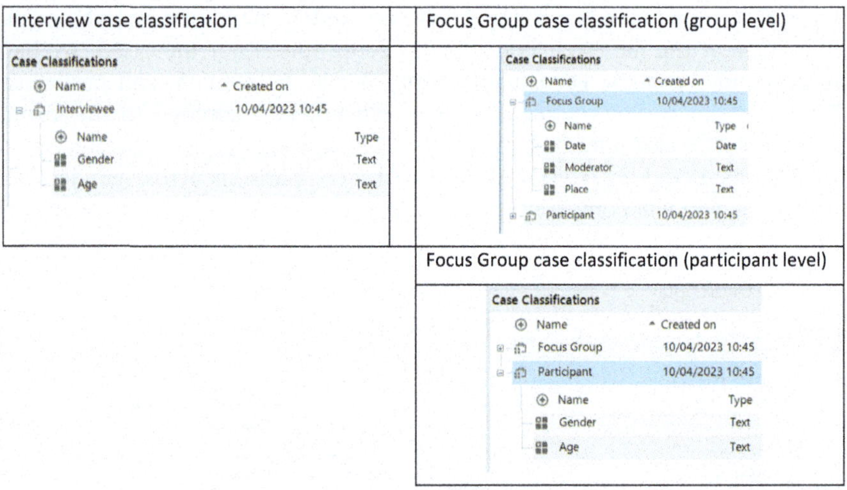

Fig. 15.4 Case classifications in focus groups—the case classification on two analytical levels

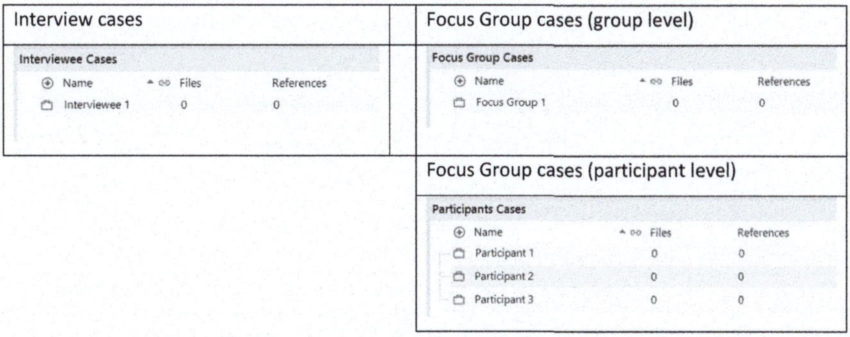

Fig. 15.5 Cases in focus groups—cases on two analytical levels

participants. Adding the case classification to the cases is not different (not shown in the figure).

The last step in working with case classifications is coding your material with the twin cases. In interview research, we explained earlier that you code the whole file (transcript) with the twin case of that interviewee. This procedure is similar to coding the focus group transcripts with the cases at the group level (see Fig. 15.6, upper right panel). For the participant level, you need to open the transcript and code each answer separately with the twin case of that participant (lower right panel). As each participant is unique within the group with different characteristics from the other participants, their individual cases must be coded each time they speak. If you have to add these case-classifications manually, this very quickly becomes a lot of (rather tedious) work. Fortunately, some solutions are available to semi-automize this coding work (see Section "Using Queries in Focus Groups" and Section "Semi-auto Coding Based on Speaker Names (in Focus Groups)" in Chap. 19) if you prepared for this classification work during the transcription.

Using Queries in Focus Groups

Working with queries is not different from focus group research in that all queries discussed in Chap. 12 are used and work similarly. Nevertheless, we show two examples in this paragraph how you can use queries specifically in a focus group context. If you do not know how to work with queries, we suggest reading Chap. 12 before you proceed.

Example 1 Coding Participants with the Text Search Query

This example cannot be done with the sample project as the interviews are formatted to work with the Autocoding function (see Section "Semi-auto coding Based on Speaker Names (in Focus Groups)" in Chap. 19). Therefore, we illustrate this example with the fabricated transcript of Fig. 15.6 (lower right panel). The prerequisite for this example is that the participant's name is

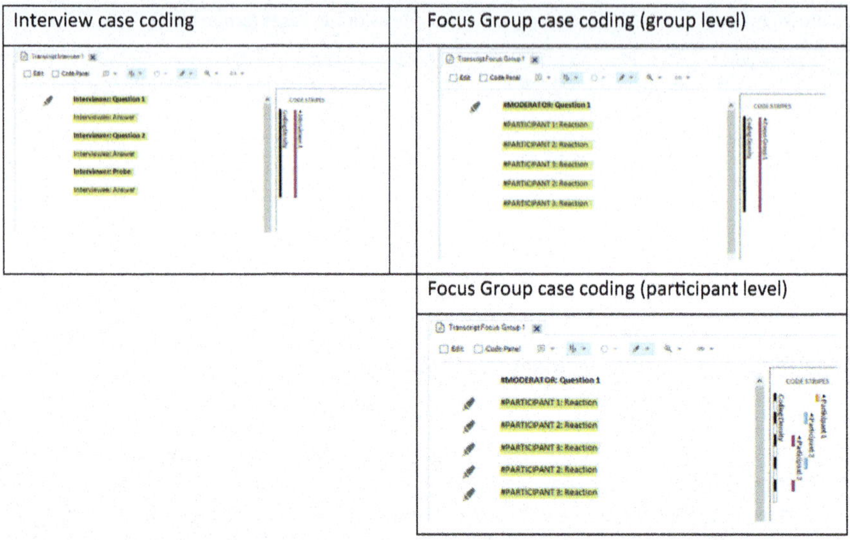

Fig. 15.6 Case coding in focus groups—case coding on the participant level

in the same paragraph as the answer. Otherwise, the option to *Spread to Broad Context* will not work. In Fig. 15.7, we show the definition of a Text search query whereby we search for "#Participant 2" to obtain all instances whereby paragraphs start with the name of that participant. We also put *Spread to Broad Context* to capture not only the name of the participant but the whole answer as well. In the screenshot, we also show the results from the query: both the participant's name and the participant's answer are highlighted in black.

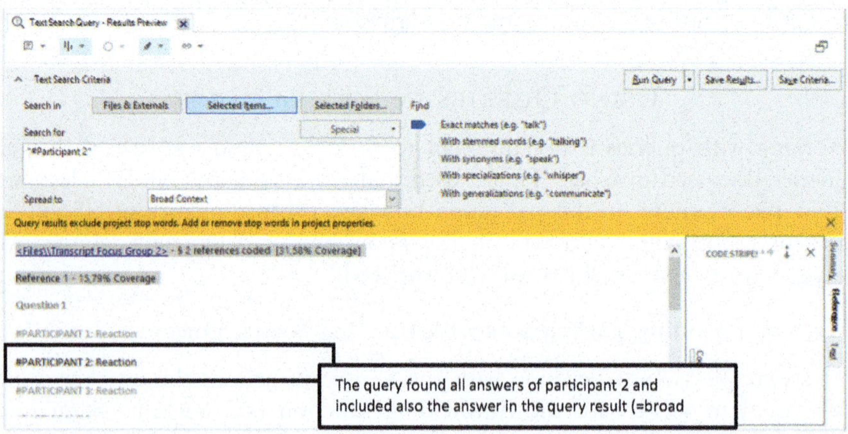

Fig. 15.7 Using the text search query to code answers of participants in focus group transcripts

When your query results show that you captured the answers of the right participant, you code the answers with the twin case of that participant. To do so, click *Save Results* and choose *Option: Create results as new code or case* or (if the twin-case already exists in your project *Option: Create results as existing code or case*. In *Location*, select the *Cases* folder to save the coded results to the correct case. To control your work, open the transcript and see at the coding stripes whether the answers of that participant are coded with the twin case.

We remind you that NVivo also has an autocoding function that automates the coding of focus group participants. We explain this tool in Section "Semi-auto coding Based on Speaker Names (in Focus Groups)" in Chap. 19.

Example 2 **Comparing participants with the Matrix Coding Query**

When analysing focus groups, the matrix coding query will be one of the main tools you can use to compare answers both at the individual and at the group level. In this example (Fig. 15.8), we show the definition of a matrix coding query where we compare content codes in the rows with cases of participants in the columns.[2] The results allow case-by-case comparisons of participants' opinions on a certain topic. When using the attributes of the individual case classification in the columns, one can also group the participants' answers according to specific background characteristics (e.g., gender or age). The query can be used on one focus group as well when you limit the scope of the query to *Selected items* and choose the focus group you want to analyse.

A final note of warning to the use of the Crosstab query. With the Crosstab query, you can add two attributes of the same classification to the columns to make an intersection of these two. Unfortunately, the attributes need to be from the *same* classifications. Hence, it is impossible to add one attribute from the group classification and one attribute from the participant classification to make a cross-level analysis. The crosstab query will not allow such analyses to be defined or executed.

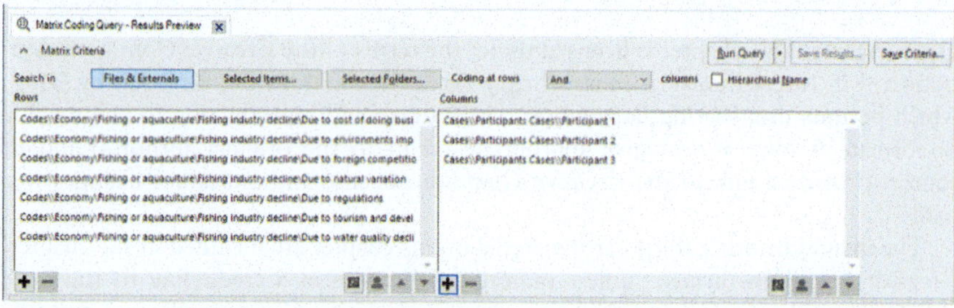

Fig. 15.8 Using the matrix coding query to compare participants in focus groups

[2] This example is a fabricated example as the sample project of NVivo does not contain any focus groups.

References

Bertrand, J. T., Brown, J. E., & Ward, V. M. (1992). Techniques for analyzing focus group data. *Evaluation Review, 16*(2), 198–209. https://doi.org/10.1177/0193841x9201600206

Carey, M. A., & Smith, M. W. (1994). Capturing the group effect in focus groups: a special concern in analysis. *Qualitative Health Research, 4*(1), 123–127. https://doi.org/10.1177/104973239400400108

Greenbaum, T. L. (1993). *The handbook for focus group research*. Lexington Books.

Krueger, R. A., & Casey, M. A. (2014). *Focus groups: a practical guide for applied research* (5th ed.). Sage.

Ledingham, J. A., & Bruning, S. D. (1999). Ten tips for better focus groups. *Public Relations Quarterly, 43*(4), 25–28.

Merton, R. K. (1987). The focused interview and focus groups. *Continuities and Discontinuities. Public Opinion Quarterly, 51*(4), 550–566. https://doi.org/10.1086/269057

Merton, R. K., Fiske, M., & Kendall, P. L. (1990). The focused interview. Free Press.

Merton, R. K., & Kendall, P. L. (1946). The focused interview. *American Journal of Sociology*, 51(6), 541–557. https://doi.org/10.1086/219886

Morgan, D. L. (1995). Why things (sometimes) go wrong in focus groups. *Qualitative Health Review, 5*(4), 516–523. https://doi.org/10.1177/104973239500500411

Parker, A., & Tritter, J. (2006). Focus group method and methodology: Current practice and recent debate. *International Journal of Research & Method in Education, 29*(1), 23–37. https://doi.org/10.1080/01406720500537304

CHAPTER 16

Using Multimedia Material (Photos, Sound and Videos)

Key messages in this chapter

- Multimedia material can be coded and queried in NVivo, like textual material.
- Pictures are most easily coded with log entries.
- Audio and video files are efficiently coded with transcription entries.
- Special tabs in the query result section show the results of queries on multimedia material.

Different or Similar?

Whether to work with multimedia material in NVivo is sometimes a choice, sometimes a necessity. In research involving interviews or focus groups, most researchers record their conversations as an audio recording. If that audio recording is digital or converted to a digital format, NVivo can import it. In that case, you have the choice of typing out a transcript of the recording or continuing to work with the original audio in NVivo.

However, if you have made observations or captured group conversations on video, the visual footage might be a dimension you want to analyse directly. Similarly, suppose you want to analyse graphic material such as advertising ads. In that case, you don't have much choice but to import this material integrally into NVivo and code it there as well.

D. Mortelmans, *Doing Qualitative Data Analysis with NVivo*, Springer Texts in Social Sciences, https://doi.org/10.1007/978-3-031-66014-6_16

If you have a choice, as a researcher, you should carefully consider whether you really want to analyse visual and audio material in its original form. If the audio or video does not add much value, then it pays off to create a (text) transcript. Large files might make an NVivo project unstable or slow, and the analysis is often smoother on text material than on audio and video sources.

In this chapter, we first illustrate how to open and view pictures, sound and video files. Next, we explain how these types of files are coded, and we end the chapter with the same examples of queries, more specifically, how you get query outputs from non-textual material.

Handling Pictures, Sound and Video

NVivo is capable of handling audio files, video files and pictures. The files are stored in the same place as the text files: *Navigation View* > **Data** > **Files**. Each type of file gets a different icon to show the user what format the file is: a drawing icon for pictures ( ), a speaker icon for audio files ( ) and a screen icon for video files ( ). As with all files, you open the file by double-clicking it. For each type of file, an extra menu is also opened: *Picture*, *Audio* and *Video*. In these menus, you find the tools to operate each file type. E.g. for audio and video files, these are the *Play/Pause* button and the *Stop* button. Also, the volume and play speed can be adjusted.

Coding Pictures

In text files or PDFs, the basic coding principle is that you make a selection of text and attach one or more codes to that selection (see Section "Inductive Coding in NVivo" in Chap. 8). This principle is kept when you want to code pictures. A selection is made in the picture to a region, and this region is then coded. When you have made the selection, you can also decide to create a log entry explaining the meaning of the selection and code that log entry. This is the overview of the coding process:

Prerequisites before you start

- The picture file must be imported into your project file and opened in the *Detail View* window.

Work plan

Step 1. Put your picture file in *Edit* mode

Step 2. Select a region on your picture
Step 3. (optional) Make a log entry from the selection
Step 4. Code the selection or the log entry.

Step 1 Put your media file in Edit mode

When you open a file in the *Detail View* window, NVivo protects the file from editing. But coding or creating log entries is considered as editing the original file. Therefore, you can only code your pictures when you first activate the edit mode. This is done by selecting the option *Edit* on the left side of the *Detail View* ribbon.

Step 2. Select a region on your picture

With the picture in *Edit* mode, you can now start making selections on the picture. Make a selection of the picture that you want to code or where you want to create a log entry (see Fig. 16.1).

Step 3. (optional) Make a log entry from the selection

As shown in the next step, you can directly code a region in your picture. But you also might be interested in labelling the region that you have selected. This is done by creating a log entry of the selection by **RMC** (*above the selection in the picture*) > **Insert Row**. The new log entry is shown to the right of your picture (Fig. 16.2). When created, NVivo shows the coordinates (pixels) of the region on the picture and allows you to label the region in the column *Content*. In the next step, you can code this log entry, which gives the same result as directly coding the selection.

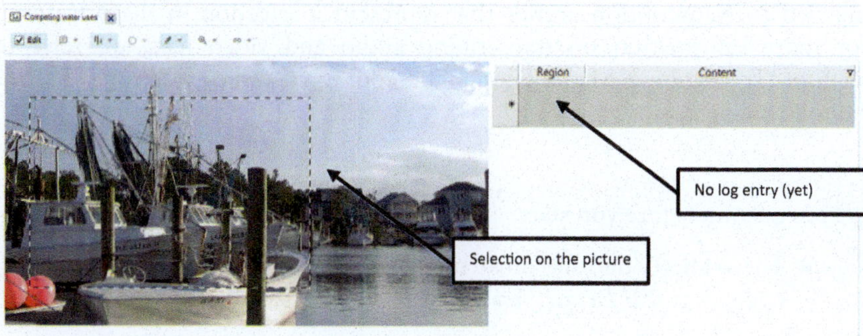

Fig. 16.1 Making a selection (region) in a picture

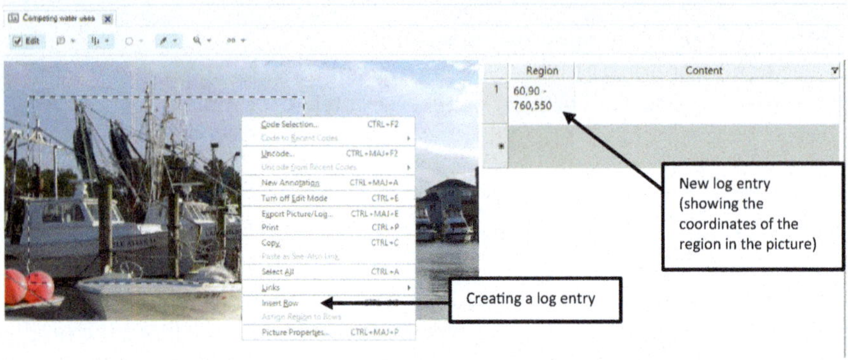

Fig. 16.2 Creating a log entry for a region in a picture

Step 4. Code the region or the log entry

The last step is to code the region. This can be done in two ways. The first way is to **RMC** (*above the selection in the picture*) > **Code selection** (Fig. 16.3, upper panel). The *Select Code Items* window appears, and you can choose the codes that must be coded to the selected region. The second way is using the log entry. When you **RMC** (*above a log entry of the picture*) > **Code selection** (Fig. 16.3, lower panel), the coding is done in a similar fashion (selecting codes and coding), but now the log entry is coded instead of the selection. We have a slight preference for this second way of coding, as using the log entries makes it easier later to find your selections back in the picture. Clicking on the log entry immediately shows the selection in the picture so that you know what region has been coded.

Coding Audio and Video Files

For audio and video files, the same basic approach will work: codes are attached to either a selection of the audio/video waveform or codes are attached to transcription entries that represent a selection in the waveform. For more information on transcribing audio and video files, we refer to Section "Transcribing Media Files". The coding of audio/video files follows the following steps:

Prerequisites before you start

- The audio/video file must be imported into your project file and opened in the *Detail View* window.

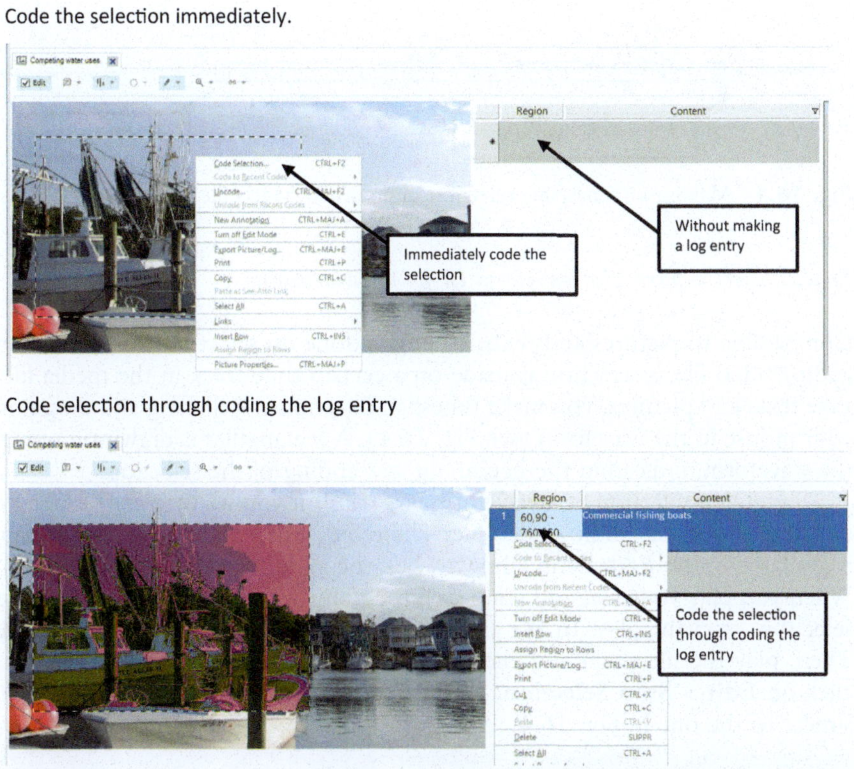

Fig. 16.3 Coding a region directly (upper panel) or through coding an entry log (lower panel)

> **Work plan**
>
> Step 1. Put your audio/video file in *Edit* mode
> Step 2. Make a selection on your audio/video file
> Step 3. (optional) Make a transcript entry from the selection
> Step 4. Code the selection or the transcript entry

Step 1 Put your media file in Edit mode

As files opened in the *Detail* View window are protected, you first need to switch on the Edit mode of your audio/video file (see the *Edit* option in the *Detail View* ribbon).

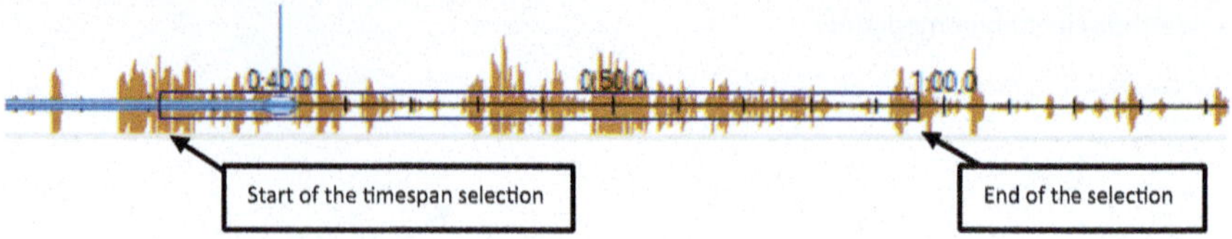

Fig. 16.4 Making a (timespan) selection in an audio/video file

Step 2. Select a part of a waveform in your video/audio

Comparable to pictures and texts, coding is done on selections of data. In an audio/video file, a selection is made on a certain time track in the media file. Like texts and pictures, you make timespan selections by clicking and dragging your mouse in the waveform (see Fig. 16.4). A rectangular is drawn on top of the waveform, indicating the beginning and ending of the selection.

Dragging a selection in a waveform is difficult if you want to start exactly a certain point in the audio/video file. Therefore, a more precise selection tool is available in the *Edit* menu (shown when the audio/video is in *Edit* mode). When you play your file, and you reach the point where you want to start the selection, stop the play. In the *Edit* menu, click on **Edit > Start Selection**. Then, play the file again and click **Edit > Pause** to halt the play. When you click on **Edit > Stop Selection,** you see that a selection rectangular is made, similar to the one in Fig. 16.4.

Step 3. (optional) Make a transcription entry from the selection

Comparable to pictures, you have two ways of coding a selection in audio/video files: directly on the selection or through a transcription entry. When you want to use the second possibility, you must create an entry when a selection was made in the audio/video file. An entry is made by **RMC** (*above the selection in the waveform*) > **Insert Row**. The new transcription entry is shown to the right of your video (or below the waveform in case of an audio file). In the first column, *Times,* the start and end times of the selection are shown. Next, you can fill in the columns *Content* and *Speaker* (Fig. 16.5).

Step 4. Code the region or the log entry

The last step is to code the timespan selection. Again, the parallel with pictures is integrated into audio/video files as well: you can either directly code the selection or code the transcription entry. The first way is to **RMC** (*above the timespan selection in the waveform*) > **Code selection** (Fig. 16.6, upper panel). The *Select Code Items* window is presented, and you can select the codes that

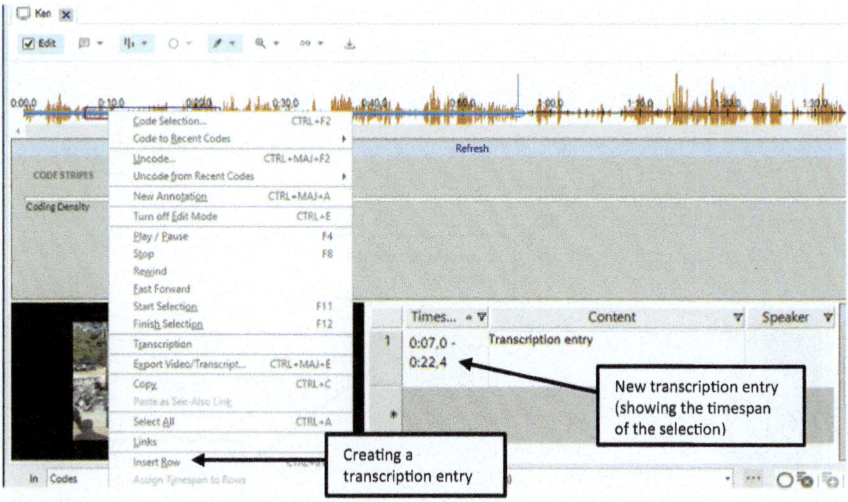

Fig. 16.5 Creating a transcription entry for a timespan selection in an audio/video file

apply to the selected audio or video reference. The second way is using the transcription entry. When you **RMC** (*above a transcription entry of the audio/video file*) > **Code selection** (Fig. 16.6, lower panel), you again need to select the codes from the *Select Code* Items window. As with pictures, we recommend this second way of coding, as using the transcription entries is easier to work with later. Clicking on any transcription entry shows the selection in the waveform so that reconstructing the selection you have made is easier.

Querying Non-textual Material

To conclude this chapter, we want to illustrate how NVivo shows results from pictures and audio/video files on the query results page. We will not repeat the creation of queries (see Chap. 12) but merely show how results from media files are adopted in the results of queries.

As an example, we made a simple Coding Query (**Explore** > **Queries** > **Coding**) and searched for the code *Fishing or Aquaculture* (child code under *Navigation View* > **Codes** > **Economy**). When we run the query, NVivo will show the textual results as the standard output. This output is located under the right-hand tab *References*. Multiple tabs are shown here, and we are interested in the tabs *Picture* and *Video* (Fig. 16.7). For pictures, NVivo selects the area on the picture that fits the query criteria. For audio/video files, the waveform with the result is selected. You can play these parts to hear/see what the result is about.

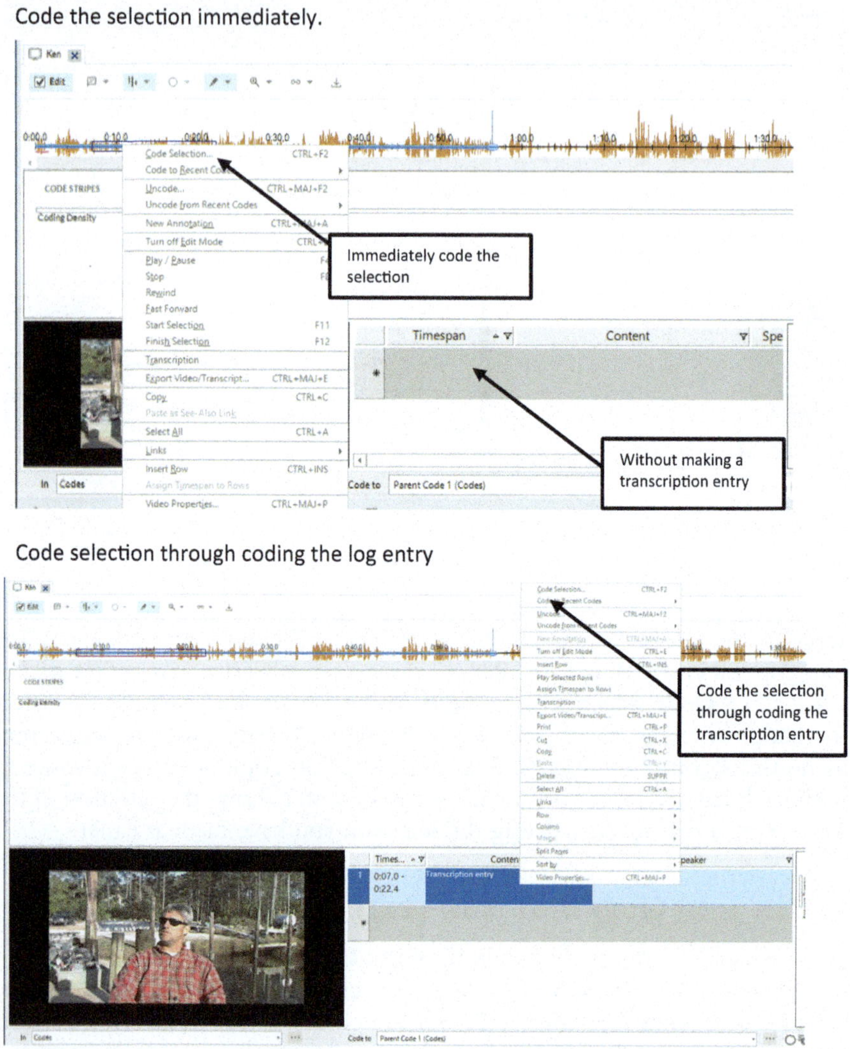

Fig. 16.6 Coding a selection directly (upper panel) or through coding an entry log (lower panel)

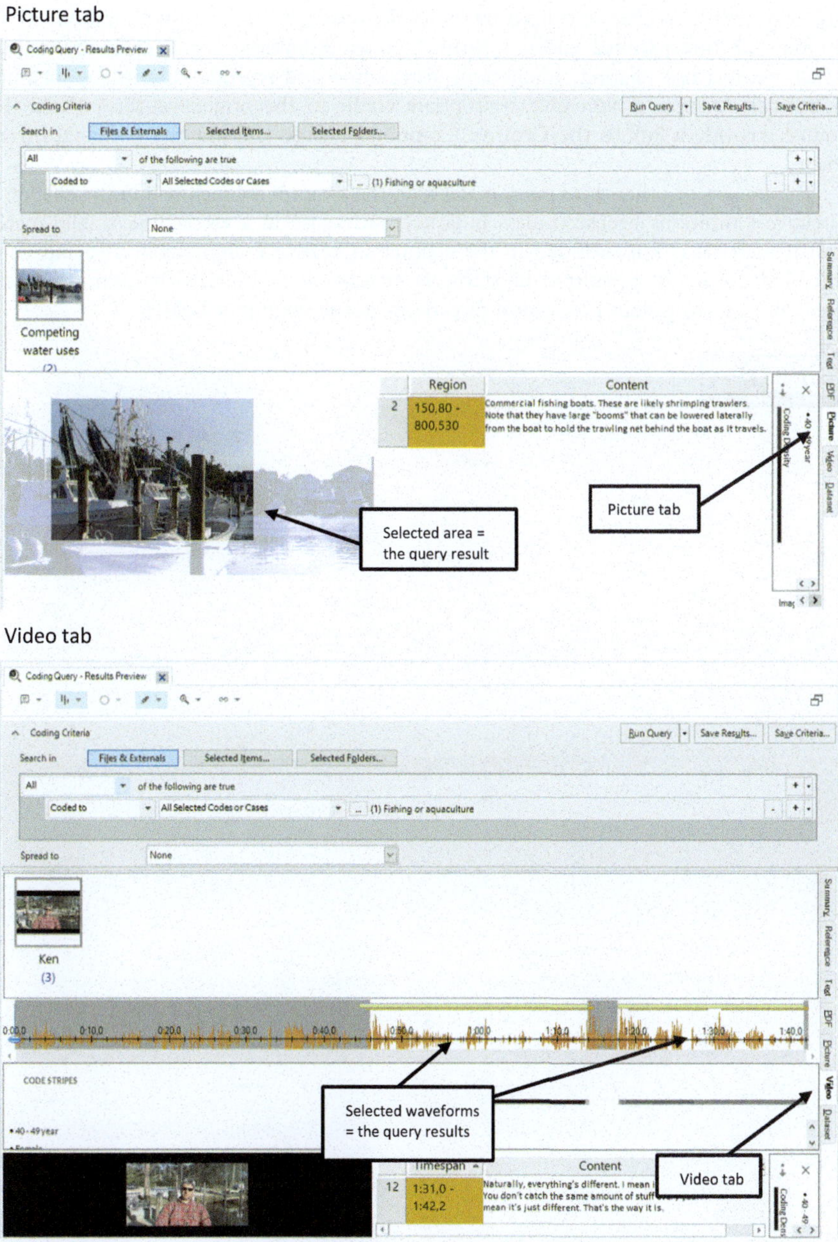

Fig. 16.7 Query results of coding query (picture tab and video tab)

CHAPTER 17

Using Social Media in NVivo

Key messages in this chapter

- NCapture is an add-in for Chrome that allows capturing online data for import in NVivo.
- First, you prepare captures in NCapture, and next, you import the captures as data in your NVivo project.
- NCapture can capture websites, X (Twitter) tweets, Facebook discussions, and YouTube videos.
- Tweets and Facebook discussions are saved in dataset project items in your project.
- Websites are saved as PDF documents in your project.

Different or Similar?

There is no agreement in the literature on what the proper definition of "social media" would entail. Still, most definitions rely on terms such as "online content" and "user-generated content" (McKenna et al., 2017). There are also many variants of social media, going from blogs over social networks to virtual social worlds (Kaplan & Haenlein, 2010). It is also clear that most social media research is quantitative in nature with big data applications, and only a smaller amount of the studies are performed in a qualitative research framework or a mixed methods design (Andreotta et al., 2019). We will not go into these different topics in this book as the actual question of what makes up

D. Mortelmans, *Doing Qualitative Data Analysis with NVivo*, Springer Texts in Social Sciences, https://doi.org/10.1007/978-3-031-66014-6_17

qualitative social media data is less relevant to this book. It is NVivo itself that defines what type of social media data can be used in the program: Twitter, Facebook and YouTube data. We, therefore, limit ourselves to explaining how NVivo deals with these types of data and forego the broader discussion in this field.

Social media data differ from interview or focus group data in combining textual and visual material (pictures or video). That being said, we have demonstrated in the previous chapters that NVivo has all the necessary tools to analyse both textual data and multimedia content. In this chapter, we will indicate the differences when working with social media data compared to the tools used in interview research or with focus groups.

Importing Social Media with NCapture

Importing social media data implies connecting to social media platforms and retrieving online discussions or videos. NVivo is not capable of doing this directly. Therefore, users can install a special add-on, NCapture, to capture social media data from their respective platforms. The plugin NCapture is currently only available for the Chrome browser. If you did not install NCapture tool as an extension for Chrome when installing NVivo, you can either look for NCapture in the 'Chrome web store' or re-run the NVivo installation file.

When installed, you see the add-on as an extension in the upper right corner of your browser window as a tiny icon ( ). If you do not see the icon, click the *extension* icon ( ) and pin the NCapture extension to your toolbar to make it permanently visible.

NCapture prepares online material before it can be imported as data in NVivo. NCapture stores the capture outside your project, and you will always need to import this capture later in your project. As a first step, we show you how NCapture works and how it captures different online sources. Later in this paragraph, we show how the transfer to NVivo projects is done. NCapture can import websites, tweets from Twitter, videos from YouTube and discussion lines from Facebook. In what follows, we show a capture of each source and explain the choices you can make when preparing these sources as data for NVivo.

Before we explain the capture, we point to the authorisation of access to sources. For Facebook and X (Twitter), you need to authorise NCapture to capture data. When you are logged in to your account and start capturing sources, you will be asked to permit NCapture to do so. For YouTube, you need to be logged in on the YouTube website for NCapture to work.

NCapture Capturing Websites

We start our explanation with websites. Even though websites are not typical social media, NCapture can also capture websites, so we start with an example of this type of data. As the sample project covers Carteret County in the US, we use https://www.carteretcountync.gov/ as the example in this paragraph to show how NCapture deals with websites. To follow the example, surf to this URL and click on the NCapture icon. The add-in shows the window in Fig. 17.1.

Before we discuss the choices to be made, a website will always be converted to a *PDF* document. This means that capturing a website implies capturing the information *at the moment* of the capture. There is no dynamic way of capturing the website whereby the information in your project will later get updated if the website changes. In your project, you will get the PDF of the website as it is shown the moment you made the capture. This caveat does not only apply to websites but to all captures that we will discuss in this chapter.

In the window (see Fig. 17.1), NCapture provides two types of captures: the full web page or only the article on the web page. The difference between these two lies in advertising. NCapture can identify advertising on the webpage. If you choose *Web Page as PDF*, you capture the whole web page, including the advertising. If you choose the option *Article as PDF*, you ask NCapture to leave out advertising (if possible) and only capture the main body of text on the website. Next, you provide the name of the capture, which will later be the name of the PDF file in your project. You can also provide a description and write a memo with the website (later converted to

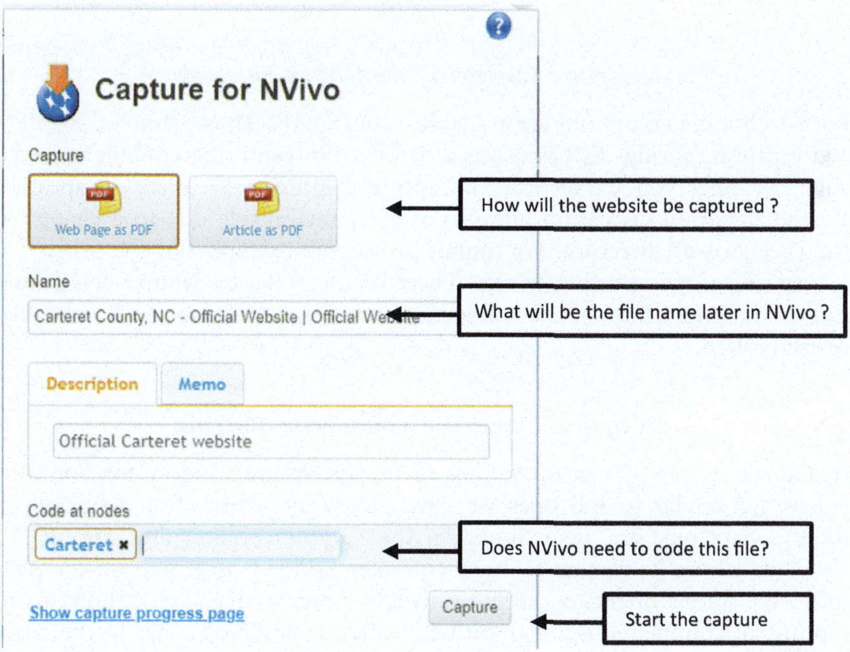

Fig. 17.1 NCapture options to capture a website

an NVivo memo). The last option concerns coding. Already, while capturing the web page, you can indicate at which codes the PDF documents need to be coded. To finalise the capture, click on *Capture*. The window disappears, but NCapture stores the capture on your hard drive. When the capture is finished, NCapture will give a pop-up message to inform you.

NCapture Capturing Tweets

In Fig. 17.2, we show the NCapture options window when you click the NCapture icon on X (Twitter) (www.twitter.com). The bottom part of the window was already explained in Section "NCapture Capturing Websites" and will not be repeated here. Different here are the options under *Capture*. The last option is irrelevant for this paragraph, as NCapture suggests treating the Twitter page as a regular web page and converting the page to a PDF. This option is offered but is not an option that treats tweets as data. The first two options (to the left) are the tweet options. Both mention a dataset as the final project item. Tweets are considered separate text, and visual messages must be kept as separate units in the project. The solution is that NVivo will put each tweet as a row in a dataset to keep the overview and facilitate the coding of the tweets (see Section "Coding Social Media Material"). Two possibilities are offered: with or without the retweets of a certain tweet. Your research question and research interests determine whether you want the tweets only or all retweets in the capture.

NCapture Capturing Content on Facebook

For Facebook, two options are available: capturing Facebook content as a PDF and capturing group discussions as a dataset. For capturing content as a web page, we refer you to Section "NCapture Capturing Websites". Capturing Facebook Group discussion lines is currently unavailable due to a change in the Facebook architecture. We cannot provide an example but the procedure is very similar to capturing tweets. Therefore, we refer to Section "NCapture Capturing Tweets" when the connection between NCapture and Facebook is re-established.

NCapture Capturing Videos from YouTube

Again, we repeat that most options in the NCapture window for YouTube videos are similar to the ones we have discussed before. The only options for this data type are again found in the *Capture* options (see Fig. 17.3). YouTube allows its viewers to leave comments on videos posted on the platform. NCapture offers to capture only the video source (left option) or to capture all comments together with the video (middle option). In the latter case, the comments are stored in a dataset in NVivo, comparable to tweets.

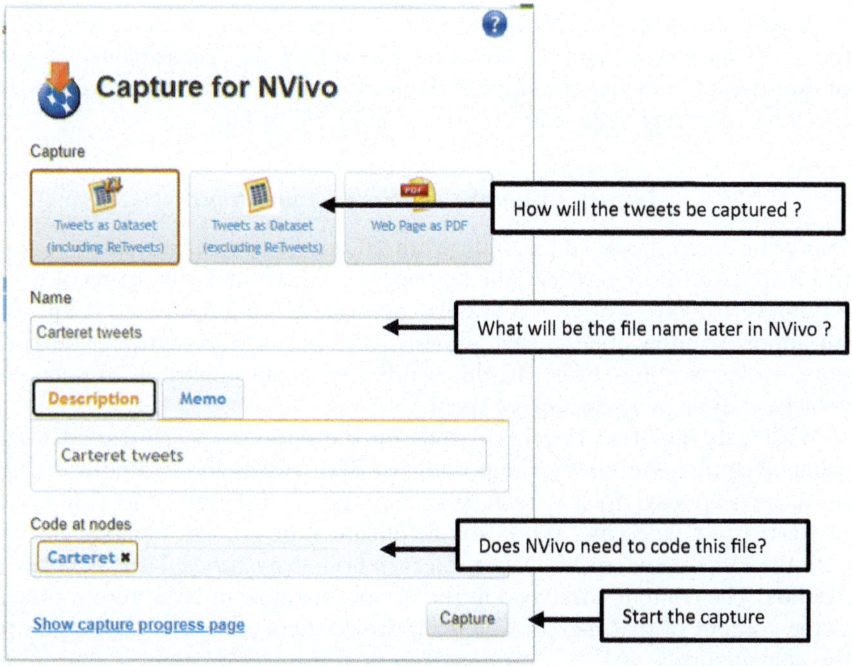

Fig. 17.2 NCapture options to capture tweets

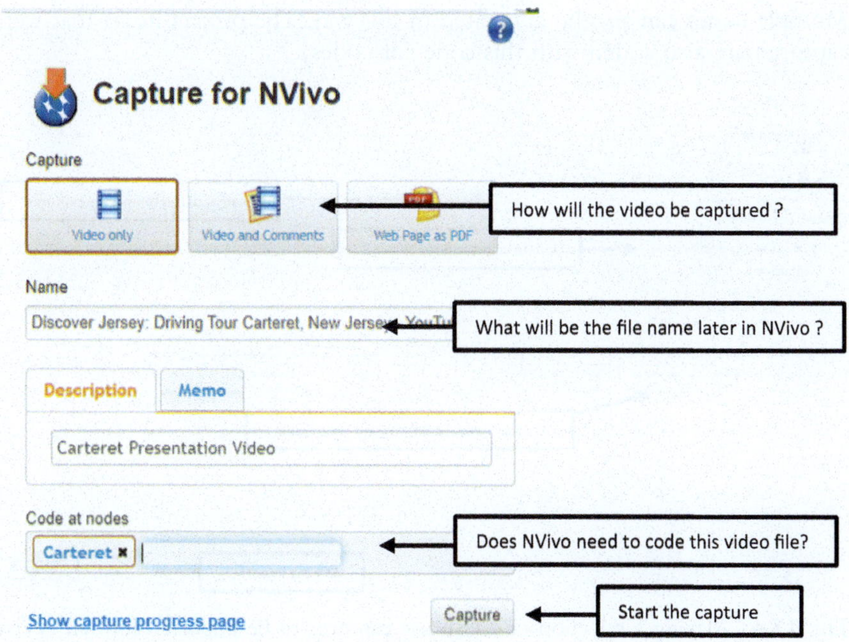

Fig. 17.3 NCapture options to capture YouTube videos

As with capturing Facebook content, YouTube is often changing the architecture of the website and the access to their videos. As a consequence, results of capturing videos may be unpredictable. Consult the Lumivero website to check whether capturing video content is available or not.

Import NCapture Entries in Your Project

Capturing social media and websites with NCapture is the first step to getting this material in your project. The second step is to import the material from NCapture to your project. To import captures, go to **Import > NCapture**. An import window appears that allows you to import the captures you have done earlier (see Fig. 17.4). In this window, you can import all captures on your hard drive or a selection of them.

What gets imported largely depends on the options you have used when using NCapture. Websites are imported as a PDF, tweets and Facebook discussions are imported to a dataset (See Fig. 17.5), and YouTube videos are converted to a video file. When you formulate a memo, the memo is created with the same name as the main project item in *Navigation View* > **Notes > Memos**. The content you typed in the options window in NCapture is placed as the content of that memo. The main project item also has a memo link to the new memo.

When you choose to have your capture coded, the new code is created in *Navigation View* > **Coding > Codes** as a parent node in the main folder. Multiple codes can also be generated in this way. The project items that were captured are also coded with this code (or codes).

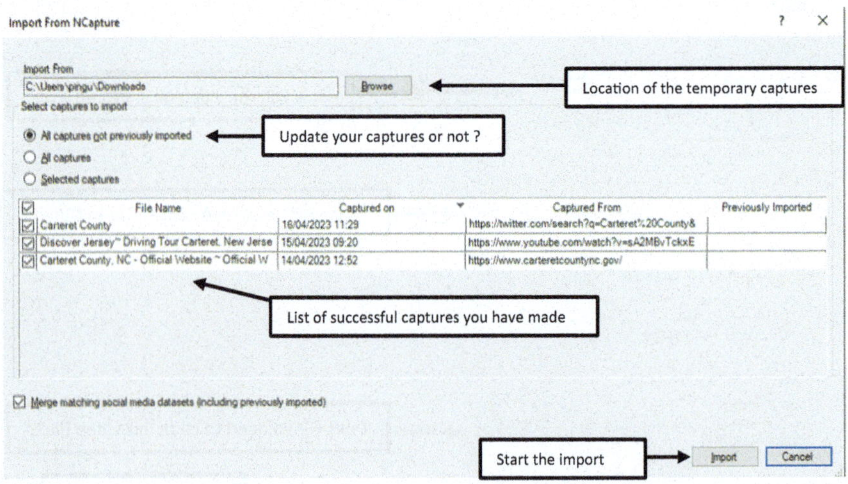

Fig. 17.4 Making a selection of NCapture captures to be imported into an NVivo project

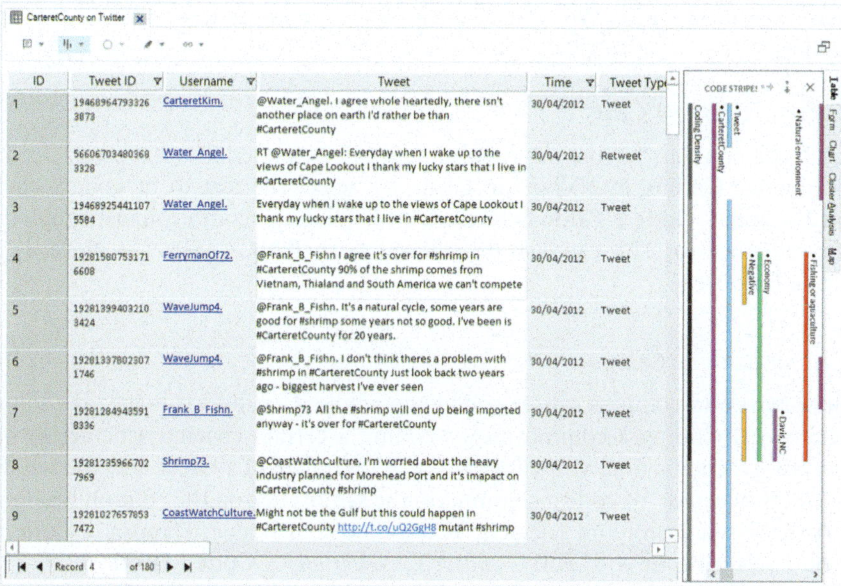

Fig. 17.5 Imported NCapture captures (tweets in a dataset project item)

When tweets or Facebook discussions contain pictures or video material, they are not imported to your project. Only the textual components of these messages are put in your project. Multimedia, however, is still available through hyperlinks. You can click on these links and see the material outside your project, but you can not code this material immediately in your project.

Coding Social Media Material

All material that gets imported as PDFs gets coded like regular textual material. We will not repeat these techniques here but refer to Chap. 8. In this paragraph, we focus on the coding of the social media datasets originating from Twitter or Facebook. As the sample project contains a sample dataset with tweets, we will use this dataset as an example in this paragraph (see *Navigation View* > **Data** > **Files** > **Social Media** > **CarteretCounty on Twitter**).

Open the dataset by double-clicking it. The dataset contains 17 columns with information. The most important columns are at the left side of the table: ID, Username, Tweet, TweetType and Hashtags. In the sample project, the dataset is already coded, so you see the coding stripes that refer to the codes attached to the data.

NVivo allows you to code the tweet dataset in three ways: coding cells (or parts of cells), coding lines and coding columns. The principles of coding data in the dataset are quite similar to the general principles of coding we have explained earlier. We refer to Chap. 8 for a comprehensive discussion on coding.

Coding (Parts of) Cells

Each cell contains textual information and can be coded like text (see Section "Inductive Coding in NVivo" in Chap. 8): select the text to be coded, click on **Dataset > Code > Code Selection,** and attach the codes you want to code the reference to. This can also be achieved with **RMC** (*above the selection*) > **Code Selection**.

Coding Rows or Columns

Both rows and columns can be coded as a whole. Coding a row means coding one tweet. Coding a column means coding a certain characteristic across all tweets. To code a line, you click on the ID number of a tweet and the whole line gets selected. To code a column, do the same: click on the column heading to get the whole column selected. The coding is analogue: **Dataset > Code > Code Selection** or with **RMC** (*above the selection*) > **Code Selection**.

Querying Social Media Material

Like coding, querying social media data is not different from querying other types of data. We refer to Chap. 12 for the general query tools in NVivo, and we also refer to the query results of non-textual material in Section "Querying Non-textual Material" in Chap. 16.

In this paragraph, we want to point to some shortcuts to available queries when you work with the dataset of tweets or Facebook discussions that NCapture creates. NVivo provides some quick ways of performing a query on your Twitter or Facebook datasets. In Table 17.1, we give a concise overview of all the shortcuts to queries you can perform when working with databases.

To conclude, we also draw your attention to the **Dataset** menu where three visualisation tools are available: Chart, Compare with (comparison diagram) and Explore diagram. These three icons give quick access to the visualisation tools discussed earlier in Chap. 10. Contrary to the query shortcuts, most of these diagrams need more specifications and are not automatically generated by clicking the icon.

Table 17.1 Quick access to queries on Twitter or Facebook databases

RMC target	*Context menu*	*Query*	*Remark*
Dataset project item	Query	Word frequency query	Automatic executing of the query with standard options
Dataset project item	Query	Text search query	Text term still needs to be selected
Dataset project item	Query	Codes coding the dataset	Gives automatically a group query with these options
Dataset project item	Query	Cases coding the dataset	Gives automatically a group query with these options
Menu when the dataset is opened in the Detail View *window*	*Icon*	*Query*	*Remark*
Dataset	Word cloud	Word frequency query	Show the word cloud visualisation of the word frequency query
Dataset	Query this dataset		All four options of the RMC are shown here (see above)

References

Andreotta, M., Nugroho, R., Hurlstone, M. J., Boschetti, F., Farrell, S., Walker, I., & Paris, C. (2019). Analyzing social media data: A mixed-methods framework combining computational and qualitative text analysis. *Behavior Research Methods, 51*(4), 1766–1781. https://doi.org/10.3758/s13428-019-01202-8

Kaplan, A. M., & Haenlein, M. (2010). Users of the world, unite! The challenges and opportunities of social media. *Business Horizons, 53*(1), 59–68. https://doi.org/10.1016/j.bushor.2009.09.003

McKenna, B., Myers, M. D., & Newman, M. (2017). Social media in qualitative research: Challenges and recommendations. *Information and Organization, 27*(2), 87–99. https://doi.org/10.1016/j.infoandorg.2017.03.001

CHAPTER 18

NVivo in Mixed Methods Studies

Key messages in this chapter

- Dataset project items contain data from surveys imported to your project.
- Datasets can be sorted and filtered on attribute items (closed questions).
- Text in open questions can be coded with the regular NVivo coding tools.
- Data (coded and non-coded) can be queried with the regular NVivo queries.

Some Theory on Surveys and Mixed Methods

It might seem strange that we insert a chapter on working with surveys in a book on NVivo. Surveys are considered to be an instrument predominantly used in quantitative research. Data from surveys are analysed with statistical software. Why would NVivo have tools to analyse quantitative data when presenting itself as Qualitative Data Analysis software? The reason has to do with the mixed methods methodology.

Often, mixed methods present their methodology as a third way (or paradigm) alongside qualitative and quantitative methods (Creswell & Plano Clark, 2017; Johnson & Onwuegbuzie, 2004; Morgan, 2007; Tashakkori & Teddlie, 2003). In this book, we do not go into the academic discussion on

D. Mortelmans, *Doing Qualitative Data Analysis with NVivo*, Springer Texts in Social Sciences, https://doi.org/10.1007/978-3-031-66014-6_18

the mixed method approach. Defining mixed methods usually starts from a general observation that it is the joint conduct of quantitative and qualitative research. Yet this simple representation is more insidious than it first appears. Indeed, Bryman (2006, 2007) indicates that when analysing published outputs of mixed methods, there is often no real integration of qualitative and quantitative research.

Within the methodological literature on mixed methods, there was an extensive search for a definition that could succinctly describe mixed methods research. The search led to a wide range of views and approaches. Johnson et al. (2007) collected many definitions from mixed methods experts and analysed them for common elements. He found that the definitions of mixed methods research had five themes in common. First, mixed method specialists all included the *merging* of qualitative and quantitative research. As said, this is how the method is widely known, and it is, above all, the primordial feature that researchers think of when talking about mixed method research. But methodologists go further. Second, they look at *where* the two methods are combined. This can happen during the research design and at the analysis, but there can also be an integration throughout the entire project. Third, the *depth* of the integration is considered. Some authors mention a mere 'merging' of the two methods, while others discuss a far-reaching integration of both. A fourth theme concerns the *way of integration*. Why do mixed methods lead to better results or deeper insights than qualitative or quantitative research separately? Reasons cited include finding more in-depth or reliable answers to research questions. Finally, Johnson finds the *orientation* of mixed methods research as a theme. This is mixed methods as a bottom-up or a top-down research method. Bottom-up would start from a researcher intrinsically interested in emancipatory and participatory research. In contrast, top-down mixed methods research starts only from the research question, leading to a mixed methods approach. Johnson's analysis shows that (1) there is no real consensus on what constitutes mixed methods research and (2) there are many more dimensions to it than simply combining qualitative and quantitative research techniques. Based on his research, Johnson suggests the following definition of mixed methods research:

> Mixed methods research is the type of research in which a researcher or team of researchers combines elements of qualitative and quantitative research approaches (e.g., use of qualitative and quantitative viewpoints, data collection, analysis, inference techniques) for the broad purposes of breadth and depth of understanding and corroboration. (Johnson et al., 2007: 123)

And how does NVivo enter this story? Even though this chapter will discuss how to work with surveys in NVivo, we want to make clear that NVivo does ***not*** have any statistical analysis tools on board. You cannot use NVivo to do the quantitative analysis of survey data. You need a proper statistical software program for that. NVivo enters a mixed methods approach concerning the text variables in a survey (e.g. open questions). When you have open questions, NVivo is well-equipped to code these questions and export the coded data back to the original survey so that you can integrate qualitative coding into your statistical analysis. You can also use queries in NVivo to explore the coded survey answers (see Section "Inductive Coding in NVivo"). The queries can combine (coded) information from interviews or focus groups with survey data. As such, you can integrate your qualitative analysis with the quantitative part. However, do not expect NVivo to have proper tabulation tools or more advanced techniques like regression or factor analysis. For those types of analysis, you need to use your statistical software.

Importing Databases

NVivo does not have a database creation tool where you can create a quantitative database from scratch. The only way to work with survey data is to import them from outside the program. In the menu **Import > Survey**, you are offered four data sources where you can import the data: Excel, Text (either txt or csv), Qualtrics, and Survey Monkey. If you use another statistical program, check the export functionality. Most programs can convert their native format to either Excel or Text. The import is done with a four-step wizard (see Fig. 18.1).

In step 1, NVivo shows an information sheet explaining what will be done during the import. Of course, the main task of the wizard is to create a database project item in your project storing the survey data. However, the information sheet shows that while importing the data, you can also partially code (autocode) the data with this wizard. Three main coding operations are shown. First, NVivo will create a case for each respondent (the lines in your dataset). NVivo will use an identification variable as the names of the cases (in step 2, a unique ID is asked to do so). Second, the closed questions in the survey are used to build a case classification and to code the cases with the respondents' responses (see more information about cases and case classifications in Chap. 9). Third, NVivo will code the open-ended questions and put the codes (name of the variable of the question) in your codebook under *Navigation View* > **Coding > Codes**. Step 3 of the wizard shows you the data you will import. This step is important to see if NVivo correctly grasps the raw data. You can also indicate how many rows the program needs to use to identify the names of the variables (column headings in this view). Step 3 asks for information to create the cases and the case classification. In the last step, you decide which variables must be imported or skipped. You do not need to import your whole quantitative database. All variables that are skipped can be indicated in the third column. In the first and second columns, you distinguish between closed and open questions. The closed question will be converted to attributes, and the values found are stored as attributes. Also, each respondent's case will get the information from these closed questions. The open and closed questions are saved as variables in your database project

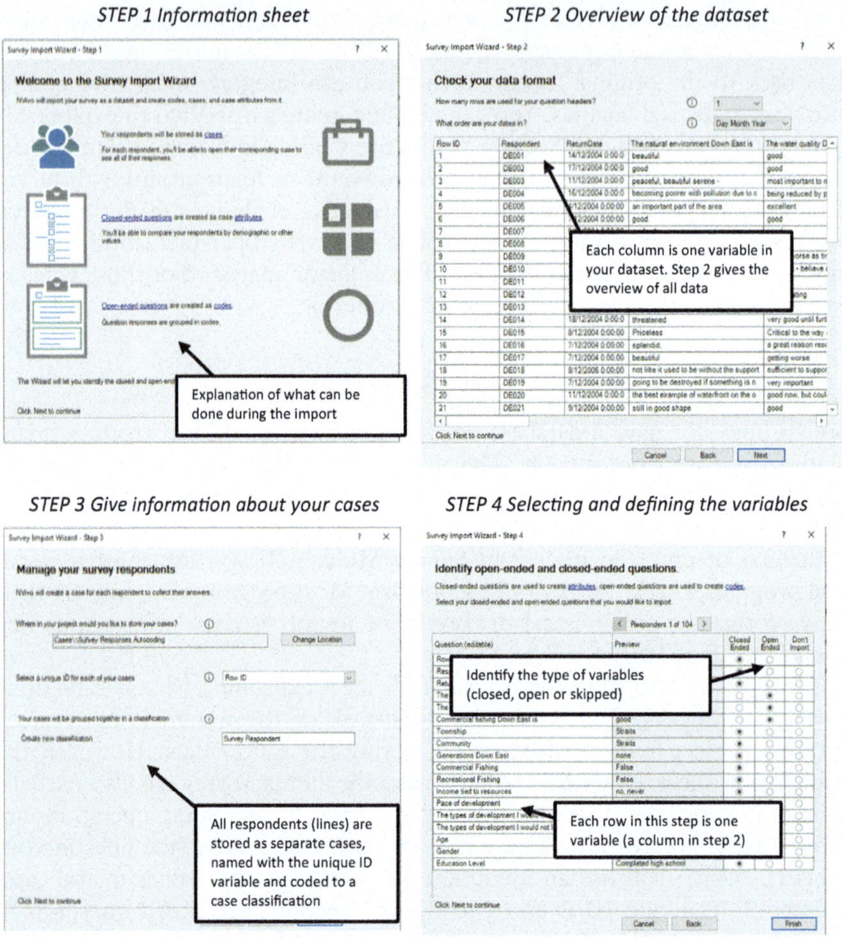

Fig. 18.1 Using the survey import wizard

item (together with the ID variable you identified in Step 3). The open question will be coded with codes named after the column heading. The whole column of a variable is coded with that code.

When you click finish, the wizard imports the data and creates a new dataset project item in *Navigation View* > **Data** > **Files**. The new cases are found under *Navigation View* > **Cases** > **Cases,** with the case classification created under *Navigation View* > **Cases** > **Case Classifications**. When you open the dataset item in the *Detail View* window, the dataset is shown with the open question having a white background (text can be selected) and the closed question having a grey background (only the whole cell can be selected) (see Fig. 18.2).

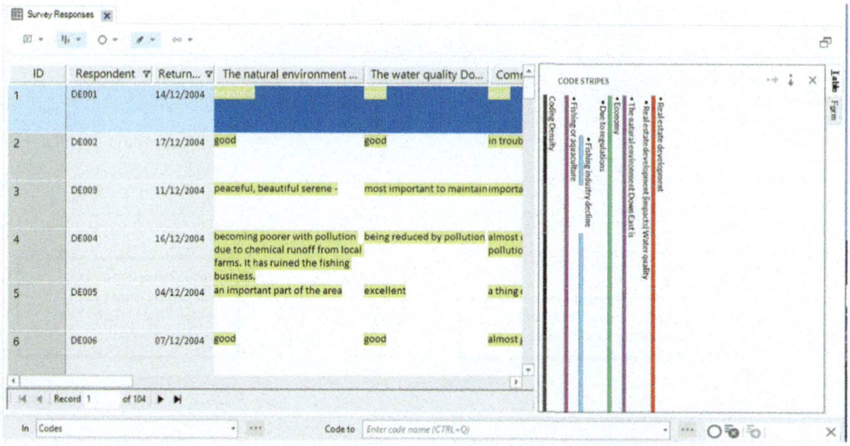

Fig. 18.2 Dataset project item opened in the detail view window

The Classify Cases from Dataset Wizard

As we explained in the previous paragraph, NVivo will create a case classification with attributes from your closed questions. Often, in surveys, you have variables with many categories (e.g. age or income). NVivo will store all these categories as separate attribute values. However, in your analysis, you should group these categories into larger classes. E.g. instead of all ages, it is better to have a limited number of age groups. To accomplish this, NVivo offers the *Classify Cases from Dataset Wizard*. You activate this wizard by RMC (above the dataset project item) > **Classify Cases from Dataset** (see Fig. 18.3).

The wizard contains four steps, but in step 4, we present two additional screenshots of the *Mapping and Grouping* window that appears when clicking on the *Map and Group* button in step 4. Step 1: explain what the wizard does. In step 2, you can create new cases for the recoded attributes. But usually, you want to add the new groupings to the existing cases and the existing classification. In step 3, you link the information of the dataset to the information of the cases. As with the import dataset wizard (see Fig. 18.1), you identify a unique ID in the dataset that is linked with the unique case names. When you use cases already created during import, it is important to select the same identification variable now in this wizard. In step 4, you select the attributes to be recoded. In the left window, you see a list of all existing attributes in your case classification. When you want to group any of them, click this attribute to the right panel. In the example, we have selected the attribute *age*. The actual definition of the new groups is hidden behind the button *Map and Group*. When clicking this button, you get the *Map and Group* window. Both the Map (left panel in Fig. 18.3) and the Group part (right panel in Fig. 18.3) have their tab in this window. You first define the *Map tab*. Here, you give the new attribute a name, and you define the characteristics of the attribute

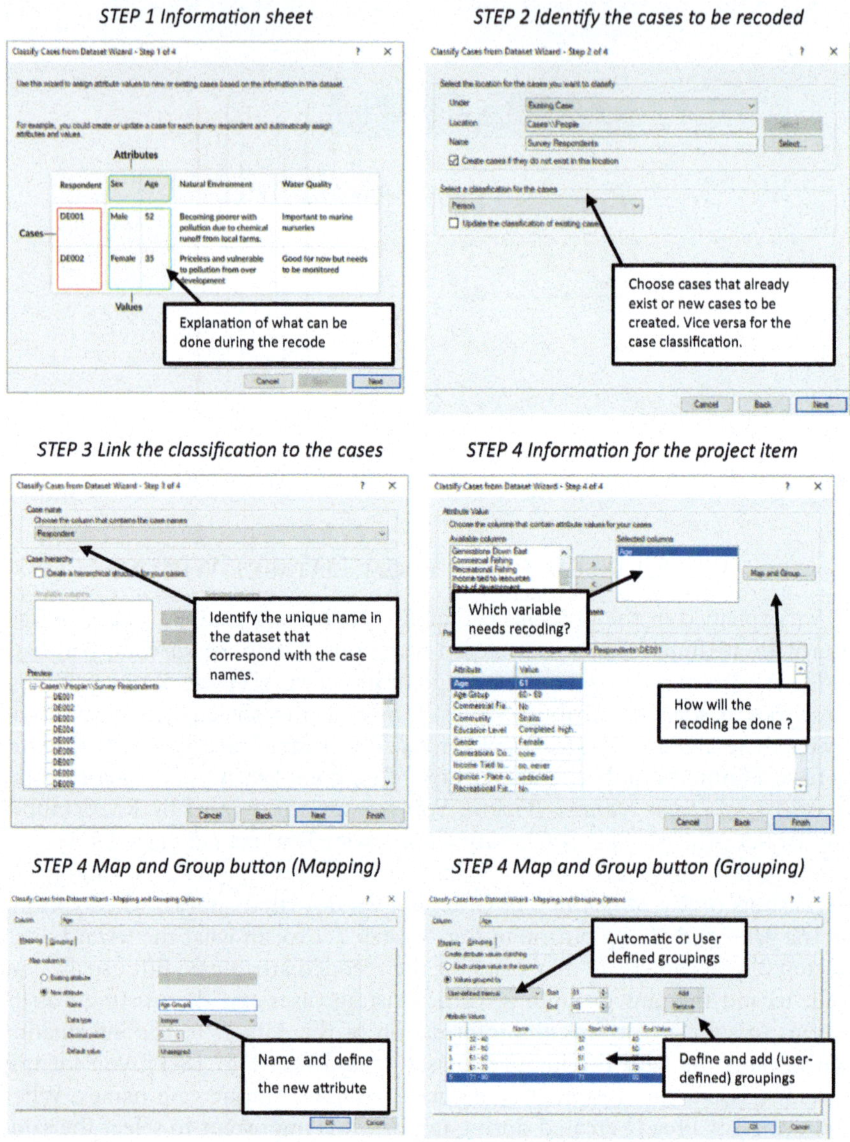

Fig. 18.3 Using the classify cases from dataset

(Data type, Decimal places, Default value). In the *Grouping tab*, you indicate how you want the new groupings to be defined. The option *Values grouped by* NVivo offers you three automatic recoding schemes: Equal Interval, Equal Area, and Standard Deviation. In the example in Fig. 18.3, we have selected the fourth option: user-defined Interval. With this option, you can manually define all categories and add them to the section *Attribute Values*. NVivo will start by giving you the lowest and highest value. With the sliders, you can create different groups and add each of them to the list. NVivo requires you to have groups for each value that the program has encountered. For example, in the sample project, ages range from 32 to 81. You cannot make groups until the age of 80 as value 81 has no group in that case. So, it would be best to define groups where all values between 32 and 81 are somehow adopted in the new attribute groups. When you finish the wizard in Step 4, the new attribute with the v-grouped values is created and saved in your classification. All cases are immediately updated and also have the correct value of the new attribute coded to them.

Sorting and Filtering Databases

An interesting database characteristic is that all information is structured in rows and columns. When you want to code the data in a dataset, it might be convenient to first sort your dataset on a certain attribute value (e.g. sort on gender to first code all answers from women and then from men) or to filter your dataset (e.g. filter on gender to only get answers from female respondents). We see this paragraph as a preparation for the coding work explained in the next paragraph.

Sorting Datasets

An NVivo dataset can only be sorted when opened in the *Detail View* window. When opened, you can get the sort window by **Dataset > Sort & Filter > Sort by Custom**. When you select one particular column, you can use the **Dataset > Sort & Filter > Sort by Column** option. When using this option, the column is automatically sorted. No user intervention is asked.

If you sort by a Custom selection, you get the *Sort Dataset* window (see Fig. 18.4). With this custom option, you can define up to three columns in your dataset on which you want the dataset sorted.

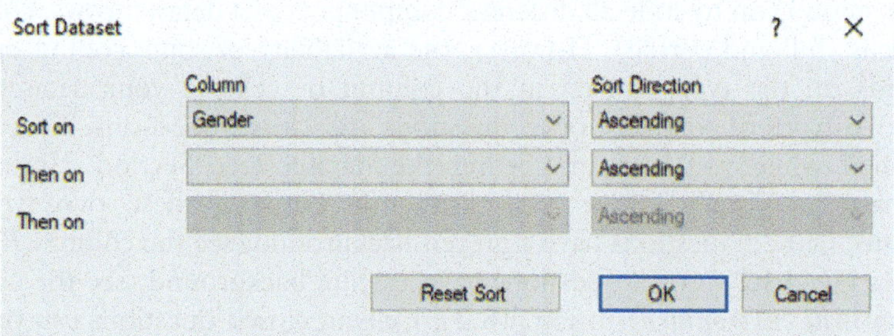

Fig. 18.4 Sorting a dataset on an attribute

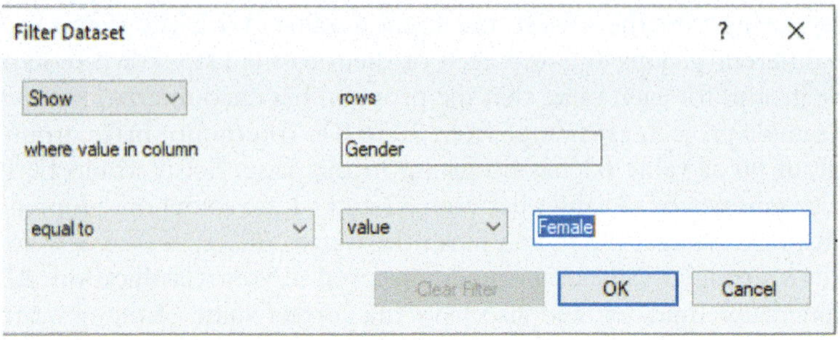

Fig. 18.5 Filtering a dataset on an attribute

When you want to remove the new sort order, you choose **Dataset > Sort & Filter > Clear all sorts** option. NVivo will then return to the default sort order (the ID variable).

Filtering Datasets

Next to sorting your dataset, it can also be very helpful to filter your dataset. Especially when you have imported a large survey, filtering respondents can help you focus the coding work you do (see next paragraph). For the filtering tool to become active, you first need to select a column in your dataset that you want to filter. In the example shown in Fig. 18.5, we first selected the column *Gender* and subsequently clicked on **Dataset > Sort & Filter > Filter Column**. NVivo shows you the *Filter Dataset* window and lets you define a selection of lines to be shown or hidden. In the example, we select all values *Female* in the column *Gender*. An important note is that you can only filter on columns with closed questions (see importing datasets) and not those with open questions.

Coding Databases

NVivo allows you to code all different components of a dataset: rows, columns and cells (full and partial). However, the availability of some coding options depends on the type of data in the column or cell. In your dataset, you have two types of columns: open questions and closed questions. These were identified while you were importing the dataset (see Section "Importing Databases"). Visually, there is a distinction between these two types of columns. Closed questions have a greyed background (see the column *Respondent* in Fig. 18.2). Open questions have a white background (see the column *The natural environment* in Fig. 18.2). Cells in closed questions can only be selected as a whole. For cells of open questions, you can only select the text within the cell (all the text or part of the text).

Table 18.1 Coding availability in components of datasets

Selection	*With RMC available?*	*With dataset menu available?*	*With small ribbon available?*
Column (open question)	Yes	Yes	Yes
Column (closed question)	No	No	No
Row	Yes	Yes	Yes
Cell (open question)	Yes	Yes	Yes
Cell (closed question)	No	No	No

Coding of data in datasets is analogous to thematic coding, as explained in Section “Inductive Coding in NVivo” in Chap. 8: you select data, and then you code the selection with the *Select Code Items* window. In that window, you can either code to existing codes or new codes. In Table 18.1, we give an overview of the availability of the coding tool, depending on your selection in the dataset.

For the cells and columns of the open questions, you can also activate the highlighting function. Coded textual material in these cells is highlighted when they are coded (either all cells if you have coded the whole column or particular cells if you have coded these cells). As closed questions can not be coded, no highlighting will be shown in that type of question. The same goes for the coding strips: coded material is taken up in the coding stripes for cell columns and coded rows.

Querying Databases

We conclude this chapter with the same message as in the previous paragraph: querying datasets is not different from querying other data types. In Chap. 12, we have elaborately explained how the different queries work in NVivo. So we refer to the information in that chapter for more general information.

As we did in the social media chapter, we limit our explanation to an overview of shortcuts to available queries when you work with the dataset project item. Several quick ways of performing a query on your dataset are available. In Table 18.2, we present an overview of all the shortcuts to queries you can perform when working with databases.

To conclude, we also draw your attention to the **Dataset** menu where three visualisation tools are available: Chart, Compare with (comparison diagram) and Explore diagram. These three icons give quick access to the visualisation tools discussed earlier in Chap. 10. Contrary to the query shortcuts, most of these diagrams need more specifications and are not automatically generated by clicking the icon.

Table 18.2 Quick access to queries on dataset project items

RMC target	*Context menu*	*Query*	*Remark*
Dataset project item	Query	Word frequency query	Automatic executing of the query with standard options
Dataset project item	Query	Text search query	Text term still needs to be selected
Dataset project item	Query	Codes coding this dataset	Gives automatically a Group Query with these options
Dataset project item	Query	Cases coding this dataset	Gives automatically a Group Query with these options
Menu when the dataset is opened in the Detail View *window*	*Icon*	*Query*	*Remark*
Dataset	Word cloud	Word frequency query	Show the word cloud visualisation of the word frequency query
Dataset	Query this dataset		All four options of the RMC are shown here (see above)

References

Bryman, A. (2006). Integrating quantitative and qualitative research: How is it done? *Qualitative Research, 6*(1), 97–113. https://doi.org/10.1177/1468794106058877

Bryman, A. (2007). Barriers to integrating quantitative and qualitative research. *Journal of Mixed Methods Research, 1*(1), 8–22. https://doi.org/10.1177/2345678906290531

Creswell, J. W., & Plano Clark, V. L. (2017). *Designing and conducting mixed methods research*. Sage.

Johnson, R. B., & Onwuegbuzie, A. J. (2004). Mixed methods research: A research paradigm whose time has come. *Educational Researcher, 33*(7), 14–26. https://doi.org/10.3102/0013189x033007014

Johnson, R. B., Onwuegbuzie, A. J., & Turner, L. A. (2007). Toward a definition of mixed methods research. *Journal of Mixed Methods Research, 1*(2), 112–133. https://doi.org/10.1177/1558689806298224

Morgan, D. L. (2007). Paradigms lost and pragmatism regained: Methodological implications of combining qualitative and quantitative methods. *Journal of Mixed Methods Research, 1*(1), 48–76. https://doi.org/10.1177/2345678906292462

Tashakkori, A., & Teddlie, C. (2003). The past and future of mixed methods research: From data triangulation to mixed model designs. In A. Tashakkori & C. Teddlie (Eds.), *Handbook of mixed methods in social and behavioral research* (pp. 671–702). Sage.

CHAPTER 19

NVivo and AI: (Semi)-Automatic Coding

Key messages in this chapter

- NVivo has both semi-automated and AI-driven automated coding tools.
- In non-AI-driven coding, NVivo uses structures already available in your data (like paragraph styles) as an aid to code your data.
- The AI-driven coding identifies themes in your data or uses previous coding work to code your data.

Artificial Intelligence (AI) and Automatic Coding in Qualitative Research

AI is everywhere. The introduction of ChatGPT brought AI to the broad public and introduced it to daily life and research. In data science, methodologists were using AI long before the introduction of ChatGPT. It is known as Text Mining (Jo, 2019; Macanovic, 2022) and is quite advanced and successfully applied to large databases of textual data. Some scholars already used the technology in qualitative research projects (Ciechanowski et al., 2020) and on a philosophical level, methodologists are discussing the implications of AI for qualitative research (Christou, 2023a, 2023b). Some AI tools were already available in older versions of NVivo but only available as an add-on for which users had to pay an additional license. From version 14, these tools are integrated into the main program at no additional cost. Therefore, we will discuss

D. Mortelmans, *Doing Qualitative Data Analysis with NVivo*, Springer Texts in Social Sciences, https://doi.org/10.1007/978-3-031-66014-6_19

in this chapter the old semi-automatic coding tools and the new AI-based coding tools. We define semi-automated coding as coding that does not use AI (text mining or machine learning) technology but merely uses some structure already present in the data and exploiting that structure to convert them into codes. When referring to autocoding tools, we refer to the AI-based tools.

Earlier in this book, we have already discussed three ways of semi-automatic coding. First, in Section "Aggregating Coding Work" in Chap. 8, we showed the In Vivo Coding technique. When selecting a word, you can automatically let NVivo create the name of a new code and code the selected reference with that new code. Second, in paragraph 13.9, we showed how to code material with queries. Most prominently, the Text Search Query is well-suited to search for textual phrases and code them afterwards. Third, we showed in Section "Importing Databases" in Chap. 18 how to autocode survey data with cases in step 3 of the *Survey Import Wizard*.

In the following paragraphs, we start with the introduction of other techniques of semi-automated coding that have not yet been discussed in Chap. 8. Later, we introduce the AI-driven tools. We end the chapter with autocoding on social media datasets.

Semi-auto Coding Based on the Paragraph Style

Paragraph styles are a collection of layout features that can be assigned to a paragraph. The term comes from word processing and is very well established in, for example, Word, where they are called *heading styles*. For example, the "Standard" paragraph style in Word includes the basic characteristics of plain text: what font will that text be in, what font size, what language, and so on. Each word processor has several built-in profiles that are very similar among themselves. For example, in addition to a "Standard" layout profile, there is usually one or more layout profiles for titles in a text, the so-called Header layout profiles. A title on the first level then gets "Heading 1" as its profile, a title on the level below it gets "Heading 2," and so on. These "heading"-styles are built-in both in Word and in NVivo.

A first way of semi-automatic coding in NVivo uses these paragraph styles in a text. The assumption behind this type of autocoding is that NVivo recognises the paragraph styles in a text and uses them as a basis to add codes to the formatted paragraphs. As a consequence, researchers must prepare their transcripts in such a way that paragraphs in their transcripts have a particular paragraph style. Moreover, the text must also be structured so that more or less the same questions have the same styles attached to them.

We clarify this with an example in Fig. 19.1 from the interview with Barbara (*Navigation View* > **Data > Files > Interviews**).

The researcher has structured the interview of Barbara with paragraph styles. Plain text has the standard profile *Normal*. The main questions (Q1 in the example) are labelled with the paragraph style *Heading 1*. Names of interviewers or interviewees have received the *Heading 2* style. All questions

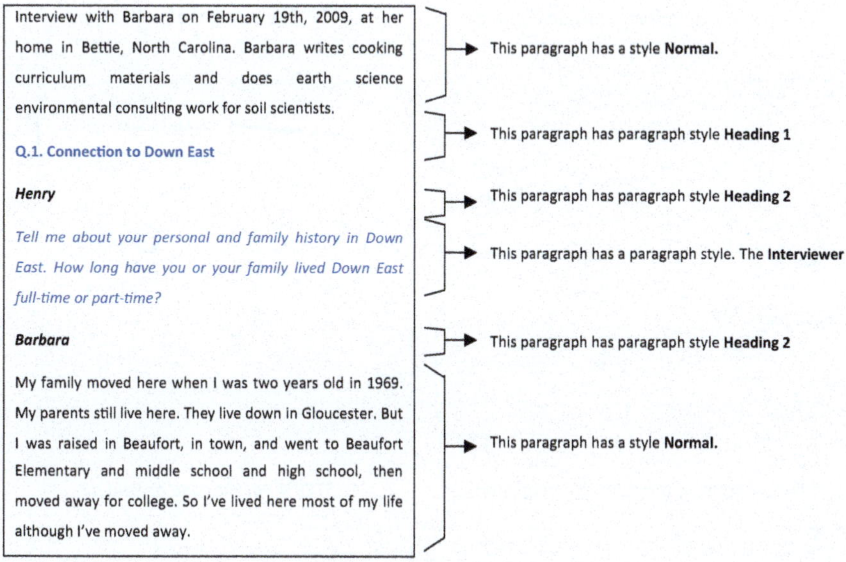

Fig. 19.1 Structuring transcripts with paragraph styles

the interviewer asks (in blue text) are marked with the style *Interviewer*. These paragraph styles can be attached to the text in Word (or another word processor) and imported into NVivo. Or you can import a transcript without any paragraph styles and add them while your document is in Edit mode.

To use the paragraph styles as an autocoding tool, you start the *Autocode Wizard* in **Document > Autocode** (Fig. 19.2). In Step 1, you determine which type of autocoding you want. For this example, we choose *the use of the style or paragraph* option. In the next step, you specify that you want the coding to be done on *Paragraph Style* in the option *Code by*. In Step 3, NVivo asks which paragraph style you want to use to determine (1) code names and (2) coding ranges. In the interview with Barbara, all questions (Q1, Q2, etcetera) have the paragraph style *Heading 1*. Therefore, we click *Heading 1* to the right-hand side in *Selected paragraph styles*. Note that you do not need to use all paragraph styles in the document. NVivo has found several styles, but we only use Heading 1 to autocode. In Step 4, you determine the location of the new codes and the grouping of these codes. Either the codes can be placed under an existing or new code that will serve as parent code for the new codes. The new codes can be placed in an existing or new folder.

In Fig. 19.3, we show the results of the Autocode Wizard. First, NVivo has created six new codes. It used the same principle as the In Vivo coding: it selects the paragraph with heading 1 and uses this paragraph integrally as the new name of the code to be created. That implies that when you use paragraph styles to code, you should use titles or short questions to format instead of complete paragraphs. If you use this type of coding with long paragraphs, your

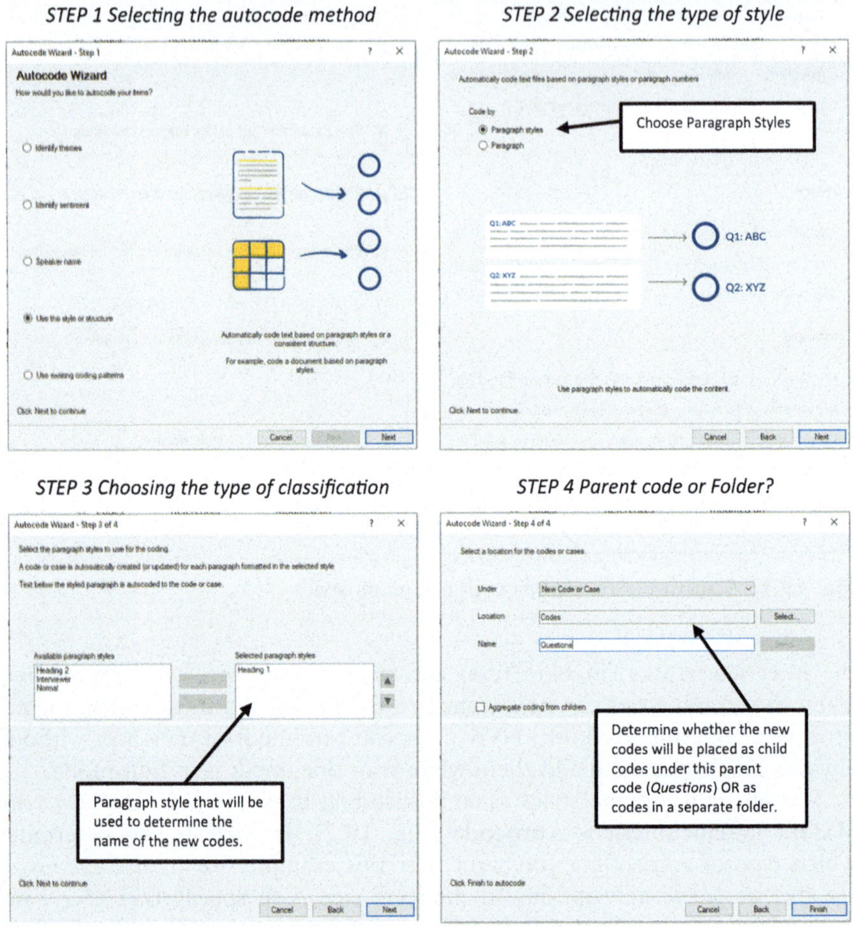

Fig. 19.2 Using the autocode wizard with paragraph styles

code names will be unworkable. In this example, all questions started with a clear indicator: Q1, Q2, etcetera. As a consequence, all new codes start with the Q1 Q2 indicators followed by the summary of the question (or, better, the question module in the questionnaire).

A second effect of auto-coding is that NVivo will immediately code your file with the new codes. The principle it uses is that all text between each instance of *Heading 1* is coded with the new code. In this example, the code "*Q.1. Connection to Down East*" is applied to all text starting from the "Q" in Q.1." until the last character before "Q.2." because there, the coding with "*Q.2. Connection to Down East natural environment*" starts. Note in Fig. 19.1 that this text has been formatted with other paragraph styles like "*Heading 2*"

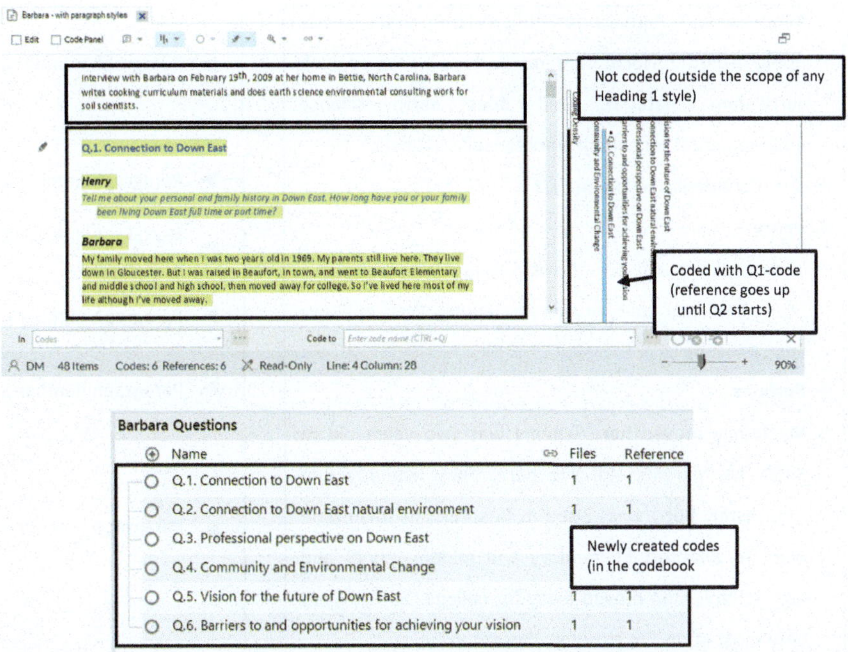

Fig. 19.3 Results of semi-auto coding with paragraph styles: document (upper panel) and codebook (lower panel)

or "*Interviewer*". None of these are used as we have not inserted them in Step 2 of the Autocode Wizard. NVivo has ignored all these paragraph styles.

Semi-Auto Coding Based on Paragraphs

A second way to code semi-automatically is based on the paragraphs in a document. You can use this when the questions are very strictly organised into paragraphs. If every first paragraph contains the answer to question 1 and every second paragraph contains the answer to question 2, then this method is useful. Again, it comes down to building your transcript so tightly that these methods work. But note that any error will be punished in NVivo.

To let a paragraph end in a text, you use the ENTER key in your word processor. The sign in Word indicates where a paragraph ends.[1] If you want to start on a new line without starting a new paragraph, you can press Shift + ENTER. In that case, the text will remain one paragraph, even if part of it starts on a new line.

We clarify this with an example in Fig. 19.1 from the interview with Barbara (*Navigation View* > **Data** > **Files** > **Interviews**). The interviews in the sample

[1] When you press the icon ¶in the menu Home in Word, you will see the paragraph endings on your screen. We made these signs visible in the text in Fig. 19.4.

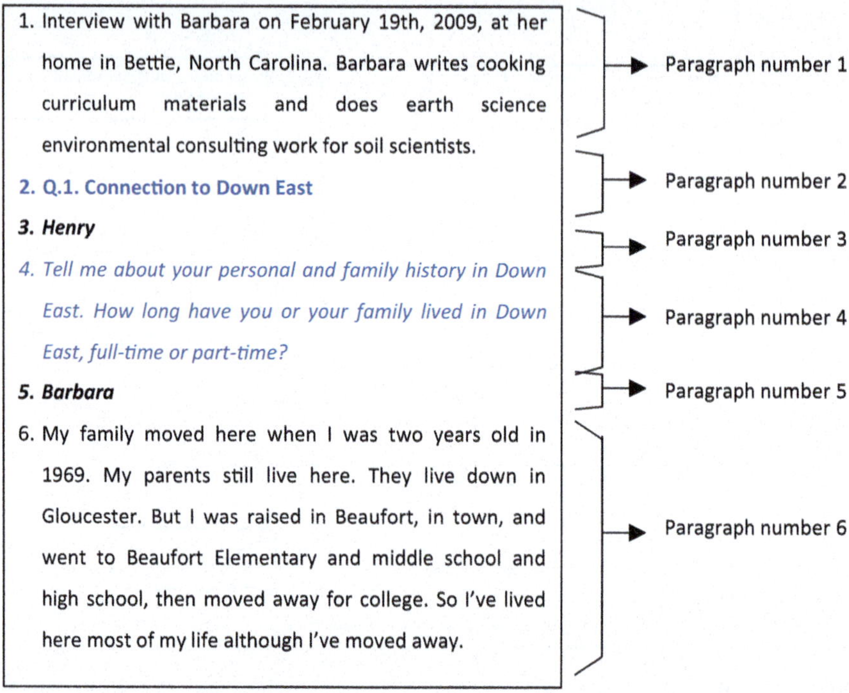

Fig. 19.4 Structuring transcripts based on the paragraphs

project have all been prepared to be autocoded with paragraph styles (see the previous paragraph). To illustrate this method of autocoding, we have numbered all paragraphs in the interview of Barbara (*Navigation View* > **Data** > **Files** > **Interviews**) (Fig. 19.4). It is not necessary to number the paragraphs, but seeing these numbers will help you understand how the Autocode Wizard will approach this text.

Autocoding with paragraphs is also done with the Autocode Wizard. You start this Wizard in **Document** > **Autocode** (Fig. 19.5). The first step is identical to the previous method: you select the option *Use the style or structure*. In Step 2, you now select *Paragraph* in the option *Code by,* as this is the method we will use in this example. In Step 3, NVivo asks for the location of the new codes and the grouping of these codes. Again, the codes can be placed under an existing or new code that will serve as a parent code for the new codes. The new codes can be placed in an existing or new folder. In this example, we have opted for the folder option as we will generate many codes (it is a long interview with many paragraphs).

To conclude the example, we again illustrate the effects in both the interview text and the codebook (upper and lower panel, respectively). In total, NVivo has created 77 new codes with names "1", "2", etcetera. Labelling of codes is now done based on the paragraph number and not on the content of

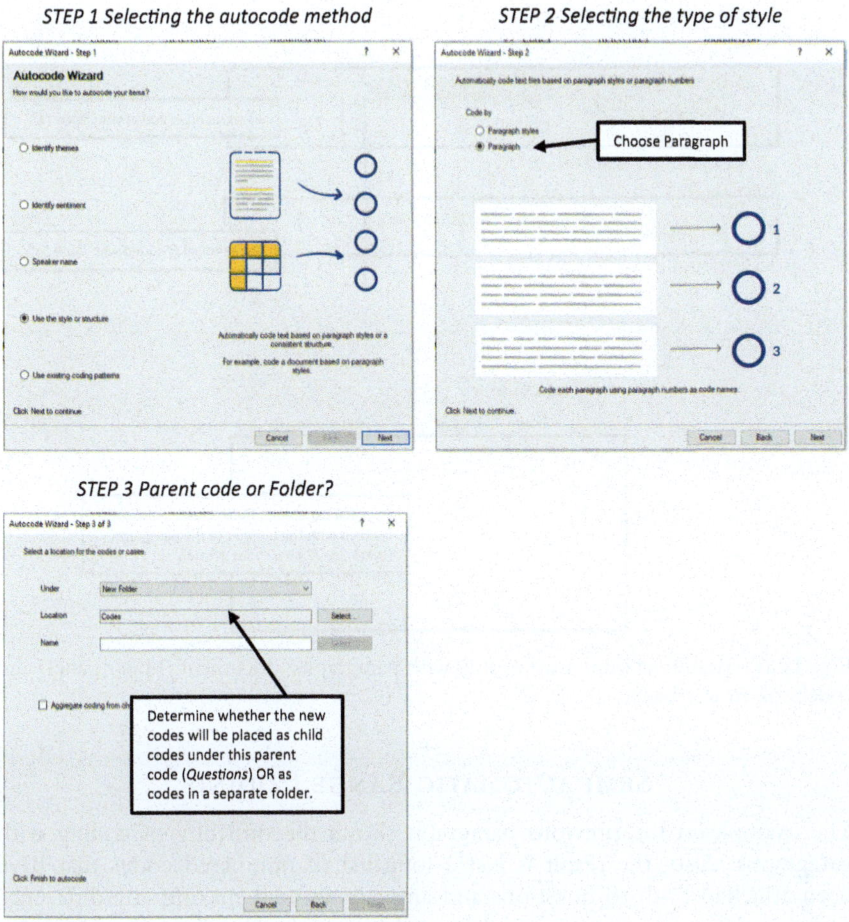

Fig. 19.5 Using the Autocode wizard with paragraphs

the paragraphs. The interview with Barbara is coded accordingly: the first paragraph is coded with the code "1", the second with code "2", and so on. As said before, these interviews were not prepared to use this type of autocoding. As a result, the coding is also not very useful. You can only use this type of autocoding if the first paragraph always refers to the same type of content, as is the case with the second paragraph, the third paragraph, and so on. As this is rarely the case, this type of autocoding will probably be used less than the one with the paragraph styles (Fig. 19.6).

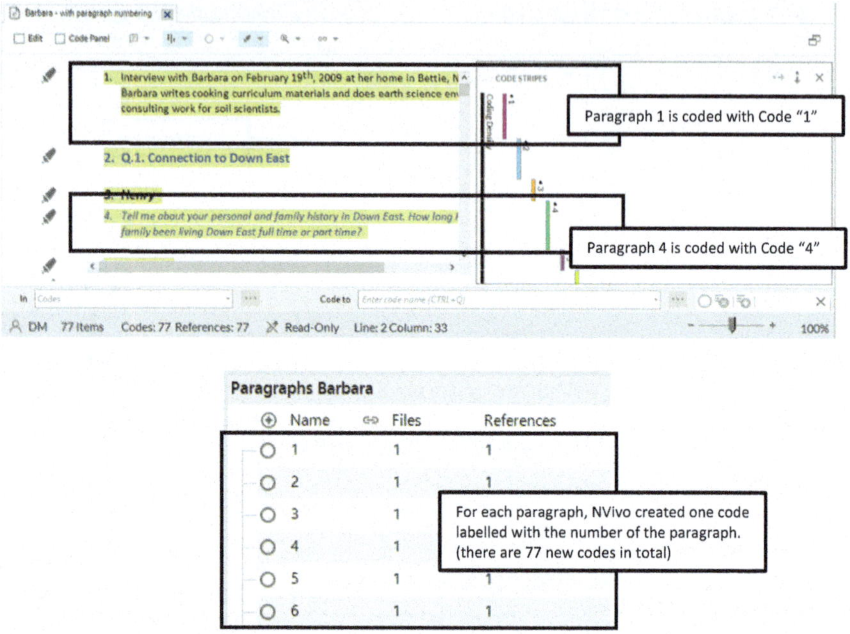

Fig. 19.6 Results of semi-auto coding with paragraphs: document (upper panel) and codebook (lower panel)

Semi-automatic Range Coding

The example in the previous paragraph shows the difficulty of coding with paragraphs. Also, the result is just a long list of numbered codes that have been attached both to questions and answers without making any difference between them. To circumvent the totally automated coding of paragraphs, Range Coding allows you to choose the paragraphs that need coding and leave out the others.

For example, we refer to Fig. 19.4, where we have numbered all paragraphs in the interview with Barbara. With range code, we can now refer to those paragraphs where the overall question and all the answers of Barbara to that question are found: paragraphs 2, 6, 10, 14, 18, 22, and 26. Activate the range coding in **Document > Range Code**. In the *Range Code* window, the paragraph numbers are entered in the option *Code* (see Fig. 19.7). In the option *Code at,* you select the code that needs to be used to code these paragraphs. Note that this time, there is no option to create a new code. You can only apply the range coding to existing codes in your codebook. We selected the code "*Q.1. Connection to Down East*" in this example, as the paragraphs all refer to that answer.

The result of the Range coding is shown in Fig. 19.8. All the paragraphs mentioned earlier are now coded to the code "*Q.1. Connection to Down East*".

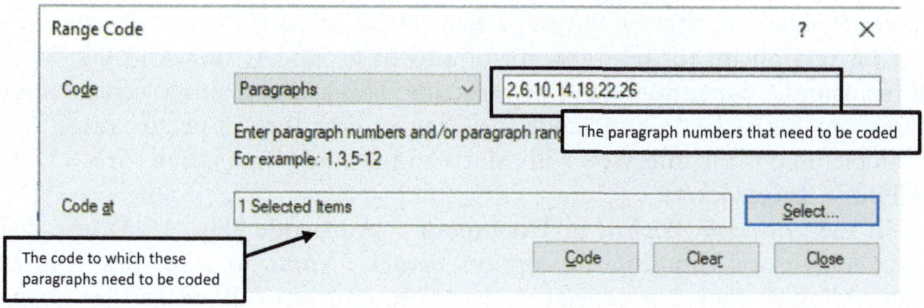

Fig. 19.7 Definition of the range coding

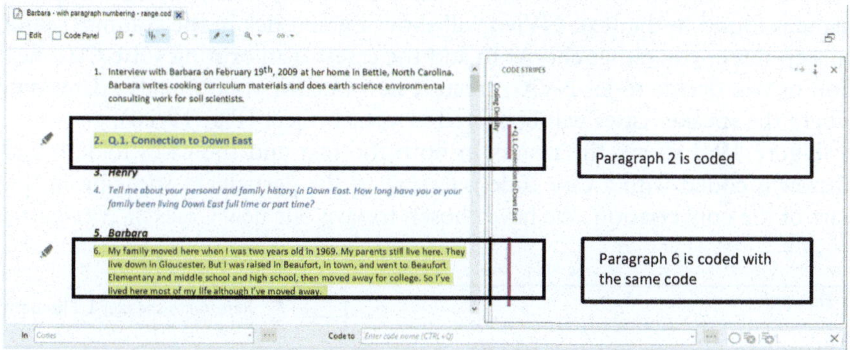

Fig. 19.8 Results of range coding

The paragraphs in between (with the names or the questions of the interviewer) are not coded. This result makes more sense than the autocoding with paragraphs. But still, you have much work identifying the right paragraphs to be coded. In this case, it might have been quicker to manually code the answers of Barbara (see Sections "Deductive Coding in NVivo and Inductive Coding in NVivo" in Chap. 8) instead of using the Range Coding.

Semi-auto Coding Based on Speaker Names (in Focus Groups)

A second way of semi-automated coding that the Autocode Wizard offers is to code speakers' names in transcripts. This is excellent for transcripts of focus groups as you have multiple participants in one focus group, and they all need their own case to be properly linked to their Case Classification. One way of obtaining this is by using the Text Search Query and code the names of participants one by one (see Section "Using Queries in Focus Groups" in Chap. 15). However, the Autocode Wizard offers a more efficient way to obtain the same result by coding Speaker Names. The example project does not have any examples of focus groups. But we will use the interview with Maria and Daniel. This interview is an interview with two participants (Maria and Daniel) which makes it a bit resemblant to the transcript of a focus group. As shown in Fig. 19.9, three names are mentioned in the transcript. This example aims to create three cases, one for each actor and code their answer with their respective cases. The example used is the interview with Maria and Daniel (*Navigation View* > **Data** > **Files** > **Interviews**).

In the *Autocode Wizard* in **Document** > **Autocode** (Fig. 19.2), the first step consists of choosing the option *Speaker Name* to start this type of autocoding. In Step 2, you need to identify the names of all speakers by adding them as rows in the upper half of the window. The lower part of the window is a control for identifying these names. With colours, NVivo shows you which part of the text will be coded with which participant. Whenever a participant's name is found in the text, NVivo will show a green tick in the column *Found*. In step three, you either choose to add the cases to an existing Case Classification or you decide to let NVivo create a new one for you. You also determine where the speaker cases will be stored in your project (Fig. 19.10).

Figure 19.3 shows the results in both the text and the cases folder. Each answer is coded with a case code referring to the participant speaking in that part of the conversation. We have chosen to save our new Cases in *Navigation*

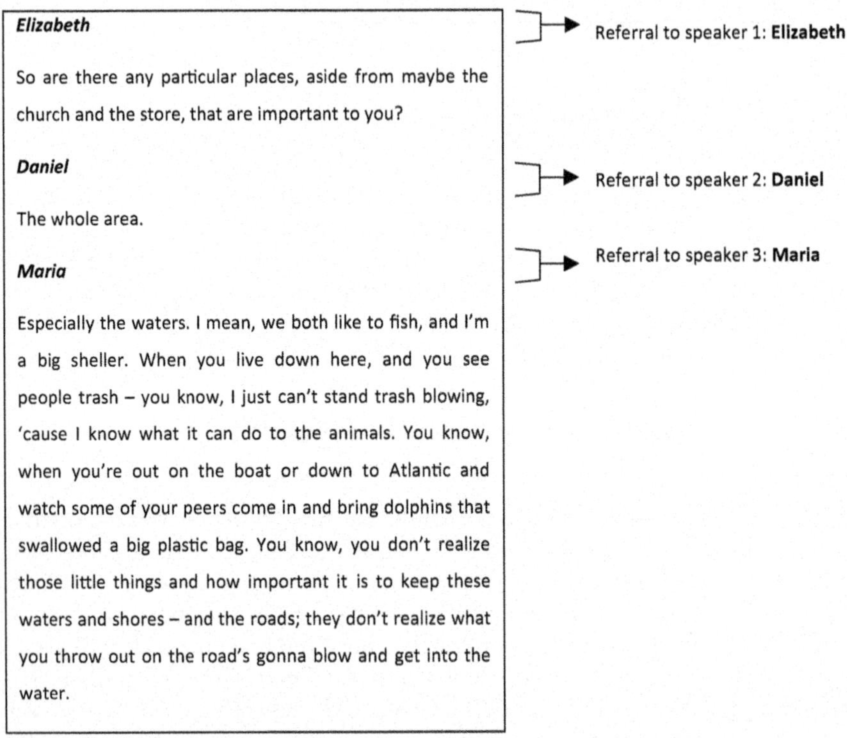

Fig. 19.9 A transcript with three speaker names (Elizabeth, Maria and Daniel)

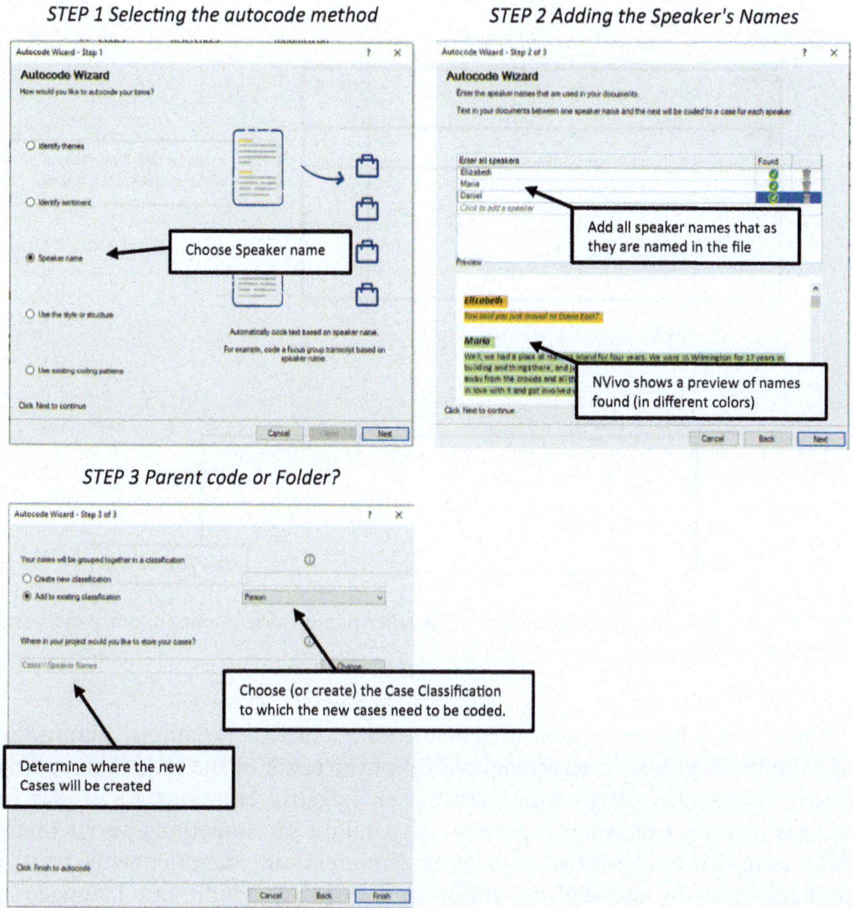

Fig. 19.10 Using the autocode wizard with speaker names

View > **Cases** > **Speaker Names**. In that folder, we find three new cases, one for each speaker. Each case is also attached to the Case Classification *Person* and the data can be filled in by asking the Case Classification Sheet (see Section "Using the Classification Sheet for Data Entry"in Chap. 9) (Fig. 19.11).

AI-Based Auto Coding of Sentiment

Sentiment coding was the first introduction of a Text Mining technique in NVivo. Sentiment coding tries to identify the sentiment (positive or negative) in the data. It is important to know that this tool looks at words and not at sentences or paragraphs (context). The coding is, therefore, done in isolation which may have as a result that you have a positive and negative sentiment coded in the same sentence. Also, NVivo is not able to recognize language

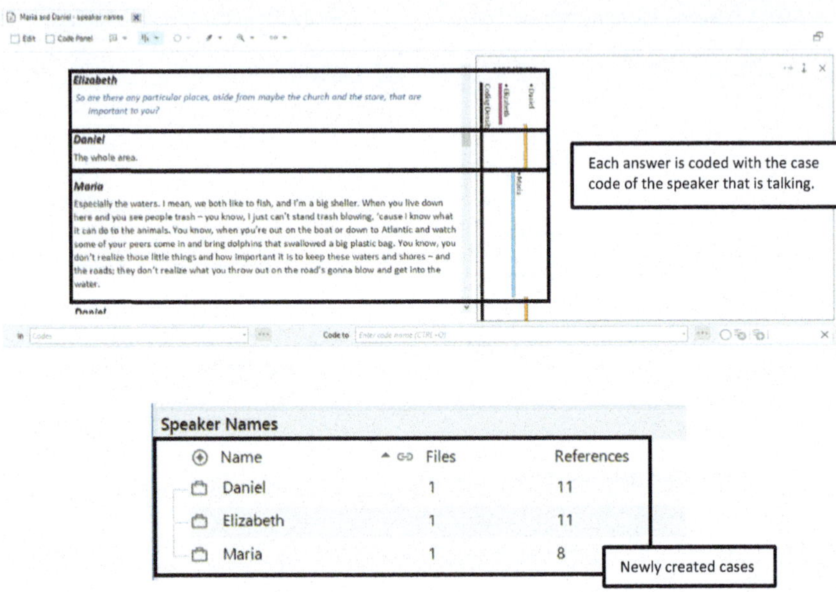

Fig. 19.11 Results of autocoding with speaker names: document (upper panel) and codebook (lower panel)

variants like sarcasm, double negatives, slang, dialect variations, idioms, or ambiguity. So you need to remain critical of the result of the sentiment coding but it can identify places where positive or negative emotions are at play or at least identify when words are used that might say something about them. Also, note that only positive or negative sentiments are identified while neutral sentences remain uncoded. In addition, it can only handle one language at a time and is limited to the standard languages for which the program has a library available: Chinese (Simplified), English, French, German, Japanese, Portuguese, and Spanish. For projects in other languages, sentiment coding will not work.

We illustrate the sentiment coding in the interview of *Barbara* (*Navigation View* > **Data** > **Files** > **Interviews**). When you open the document of this interview, you can start the Autocode wizard from **Document** > **Autocode** (see Fig. 19.12). In the second step, NVivo only asks you what the context is that needs to be coded. You can choose between sentences or paragraphs. Clicking on finish immediately starts the coding process.

When the coding is finished, NVivo shows you two visualisations (see upper panel in Fig. 19.13): a matrix with the source(s) in the rows and the sentiments in the columns. Next, a hierarchy chart is shown that represents the weight of each sentiment by the size of rectangles in a box. Sentiments are coded to your data. That is visible when you open the *Coding Strips* in the interview with Barbara. However, in your codebook, no sentiment codes are shown. The

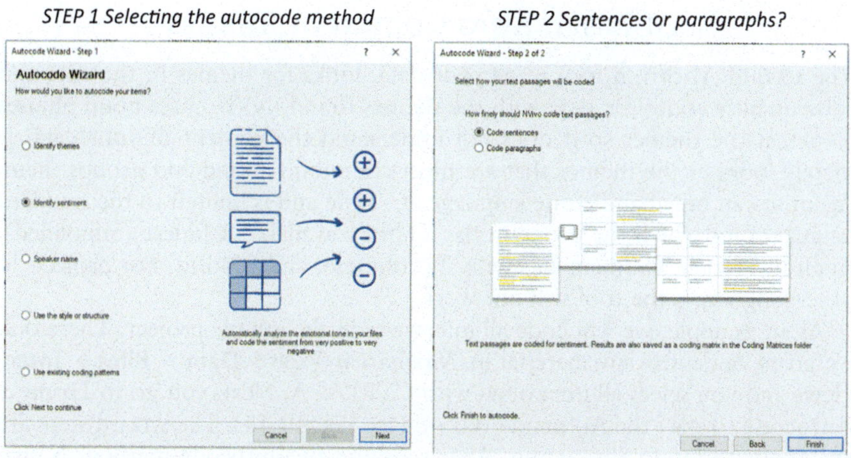

Fig. 19.12 Using the autocode wizard to code the sentiment of your data

sentiment codes are fixed (named) codes and can be found in (*Navigation View*) > **Sentiment**. The structure of the codebook contains two parent codes (Positive and Negative), with two child codes identifying the intensity of the emotion underneath.

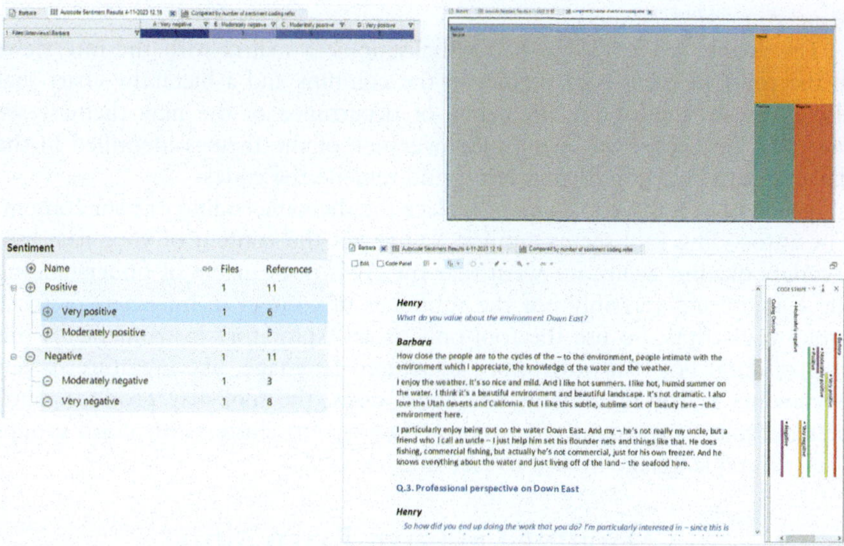

Fig. 19.13 Results of autocoding sentiment: visualisations (upper panel) and coding (lower panel)

AI-Based Auto Coding of Themes

The second AI-driven tool scans your data, looks for themes in the data and subsequently codes the data with the themes found. NVivo uses noun phrases to detect the themes so it does not understand the content of your data. It merely looks at the themes that are most commonly found and groups them. Again, it can only handle one language at a time and is limited to the standard languages for which the program has a library available: Chinese (Simplified), English, French, German, Japanese, Portuguese, and Spanish. For projects in other languages, the tool will not work.

As an example, we will code all interviews in the sample project. Therefore, we go to our interview material in *Navigation View* > **Data** > **Files** > **Interviews** and you select all interviews with CTRL + A. Next, you go to **Home > Autocode** to start the Autocode Wizard (see Fig. 19.14). The wizard steps are analogous to sentiment coding. After selecting the autocode method, NVivo immediately starts identifying themes (in some cases the program needs to install a language module but that only takes a few seconds). The second step presents an overview of the themes identified in this data. You can review them and unselect themes that you do not want to create in your project. In the third step, you determine the coding context: paragraphs or sentences (or cells in case you work on a dataset). The last step asks for the place in (*Navigation View*) > **Coding > Codes** where you want to save the codes from the themes. They will all be placed under one parent code (NVivo suggests the name *Autocoded Themes*).

The results are shown in two visualisations: a matrix with the interviews in the rows and the main themes in the columns and a hierarchy chart that also gives an idea of the frequency of occurrence of the new themes (see Fig. 19.15). These two give a nice overview of the themes identified in the data and can help you further refine and rename the codes.

A final caveat: this tool can not replace the thematic coding of your content. The autocoding tool seems a quick way to get the content of your data into a codebook. But again, we warn that the AI is not capable of understanding the content and certainly not the subtleties of content in interview or focus group transcripts. So use the tool only as an exploratory instrument for this type of data. For data with more structured language, like literature, policy documents or law documents like court cases, the tool may produce much better results as standardized language is easier to cluster with noun groups than natural language in face-to-face interviews.

Machine Learning Based Auto Coding of Coding Patterns

The third AI-driven tool works with machine learning (Pinheiro & Patetta, 2021). Machine learning differs from automated theme detection in that it needs training data to learn from to perform its actual task (coding). In NVivo,

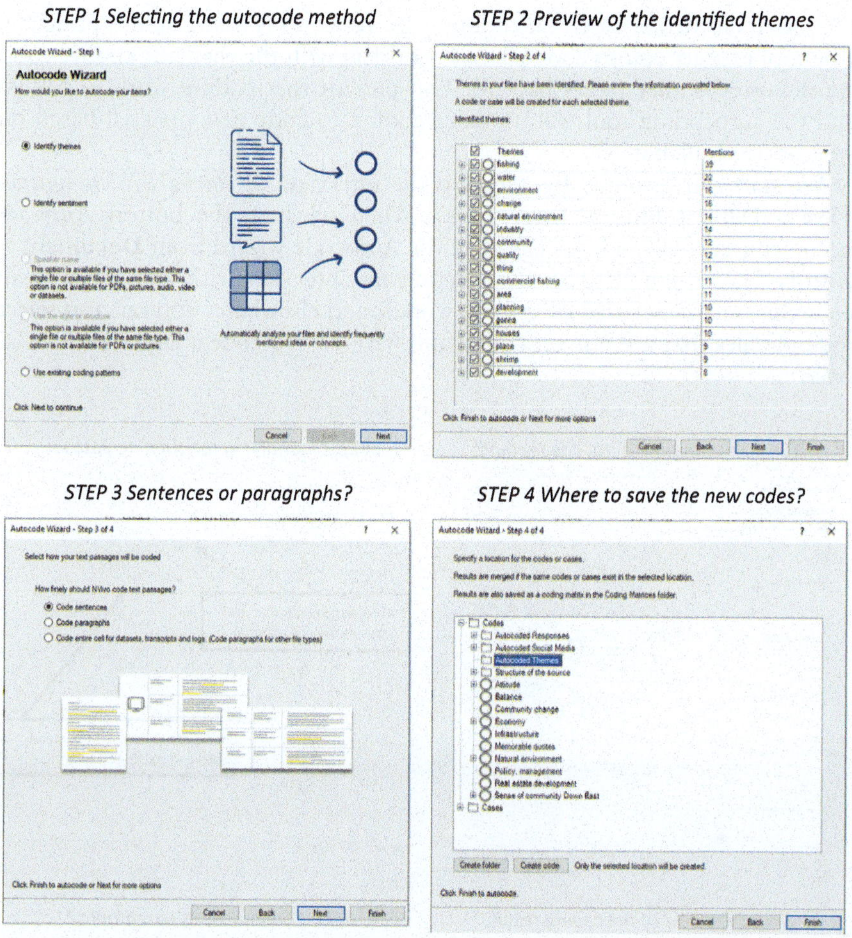

Fig. 19.14 Using the autocode wizard to identify themes in your data

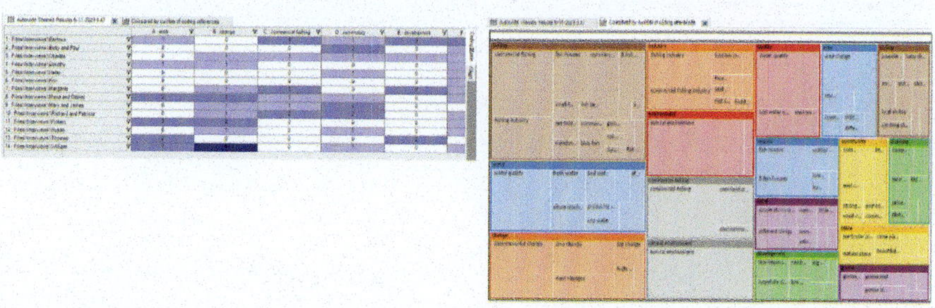

Fig. 19.15 Results of autocoding for themes: visualisations

the tool to autocode based on coding patterns works is based on this principle. The tool expects you to code part of your data with the thematic coding tools we discussed earlier (see Chap. 8). You present that coding work to NVivo and the autocoding tool will use your coding to code new material based on what it has available.

We start our example by selecting the interview of *Barbara* (*Navigation View* > **Data** > **Files** > **Interviews**). When selected, the button *Autocode* becomes accessible and you can start the Autocode wizard from **Document** > **Autocode** (see Fig. 19.16). By selecting the interview with Barbara, we will code this interview based on coding work found elsewhere. You can also select multiple files that are not yet coded and have NVivo code a larger number at once.

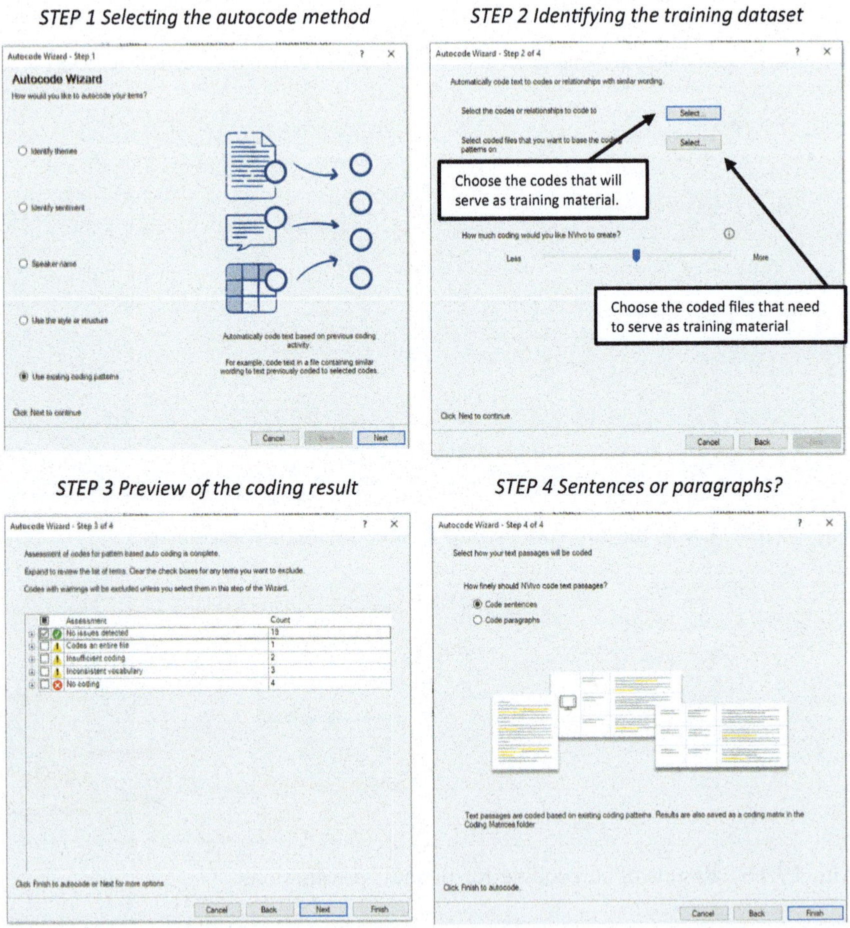

Fig. 19.16 Using the autocode wizard to code on existing coding patterns

Fig. 19.17 Result of autocoding for existing patterns

Step 2 is the most important step in this wizard. Here, you identify the training dataset. Two selections need to be made (see Fig. 19.16). First, you need to select the codes from your codebook that will serve as example codes for the machine learning. Second, you need to select the files that have been coded with these codes to identify the training dataset NVivo can use to learn how your codes were applied to actual data. A slider in the middle of the window asks how much coding you want to procedure. The choice "less" implies a higher threshold to identify "same coding reference" while the side "More" lowers that threshold and allows the machine learning to find similarities in the data more easily.

The third step in the wizard gives feedback on the machine learning. You gain insight into which codes could be used to code new material (*No issues detected*) and which codes were more problematic. When reviewing the result, you might go back one step and move the slider more to the left or the right to change the threshold of similarity (Fig. 19.17).

The result of the procedure is only a Matrix in this case. In the rows, you get the files that were coded. In the columns, you get an overview of the (existing) codes that were applied to this new material. By double-clicking on a cell, you get the references that were coded with the codes.

A caveat: this procedure is more refined than the autocoding for themes. Here the machine learning actually uses your scholarly input to do autocoding. But again, machine learning does not understand the new data and only looks for similar patterns in the data. When similar references get coded, they might be erroneous because the interviewee was sarcastic and meant the exact opposite of what was said. The autocoding tool will not recognize this and just code the reference as similar to earlier references. Keep a critical attitude towards the coding that is done with this procedure to avoid erroneous conclusions in your research.

Semi-auto Coding Social Media and Surveys

In Chap. 17, we discussed how you work with Social media data. One of the possibilities of importing social media is to divert the social media data to a dataset where all data are, for example, tweets. In Chap. 18, we discussed working with dataset project items. As these project items are similar (they are both considered dataset project items), we will discuss the semi-autocoding of dataset project items in this paragraph. You can also use the

AI-generated autocoding tools which we will not repeat here. They work similarly as discussed in the previous paragraphs. Only the semi-automatic coding based on the database structure differs from what we discussed in Sections "Semi-auto Coding Based on the Paragraph Style" and "Semi-auto Coding Based on Paragraphs".

As an example, we use the tweets on Carteret County, stored in a dataset at *Navigation View* > **Data** > **Files** > **Social Media**. We developed the example on a dataset with social media, but the explanation is completely analogous to datasets containing survey data. Therefore, we only present the autocoding once and apply it to a dataset with tweets.

When you open the dataset, we can start the Autocode Wizard at **Dataset** > **Autocode**. In the first step, you see all AI-based tools. But for this example, we choose the semiautomatic tool *Use the style or structure* (Fig. 19.18).

In Step 2, NVivo presents three ways of coding the data in the dataset. We show each way in the screenshots below with a short explanation. Note that in the screenshots, NVivo illustrates with schemes what will be coded (yellow = coded text) and with which codes (red area Code). These schemes perfectly illustrate the choice you make in your autocoding.

Autocode Wizard CHOICE 1: Code to codes or cases for each value in predefined Twitter columns

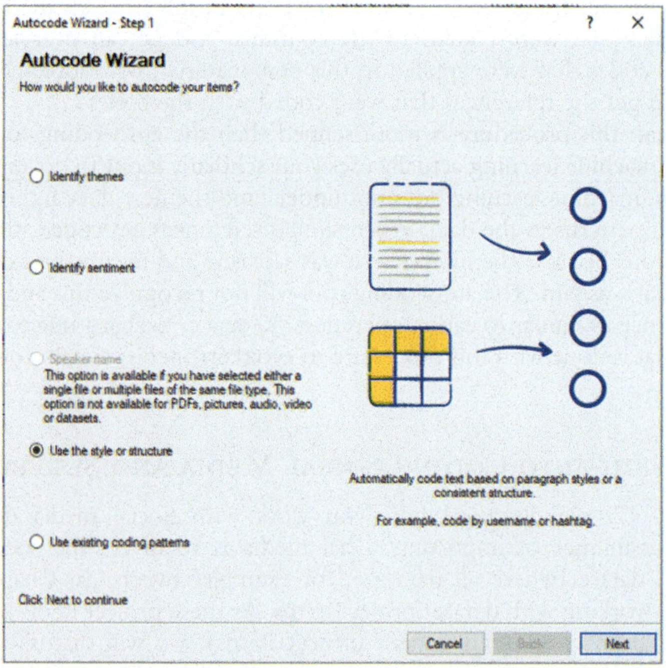

Fig. 19.18 Autocode wizard—start screen

In the first choice (Fig. 19.19), you work *within* the rows of the dataset. NVivo will create new codes based on the content of one of the fields in the database. These codes are used to code text in another field.

In step 3 of the wizard (not shown in the figure), you typically use the text in the variable "Hashtags" (or closed questions in a dataset with survey data) as new codes and apply these codes to the text of the variable "Tweet". Other possible fields to be used as codes are the variables "Mention", "Tweet type", and "Location". At the same time, NVivo will create new cases for each data line (i.e. each tweet) using the variable "Username" as the name of the case. The variable "username" can only be used to create cases and is not available to be selected for the coding part.

In Step 4 (not shown in the figure), you determine the name and location of both the new (parent) codes (or folder if you wish) and the new cases (can also be a folder).

Autocode Wizard CHOICE 2: Code columns

The second choice allows you to code all the fields in a particular column in the dataset with the name of that column (see Fig. 19.20). In step 3 (not shown), NVivo asks the user to select the column that should be coded. Typically, this is the column of the variable "Tweet", as this contains the content of the tweets (or an open question when you autocode surveys). In step 4 (not

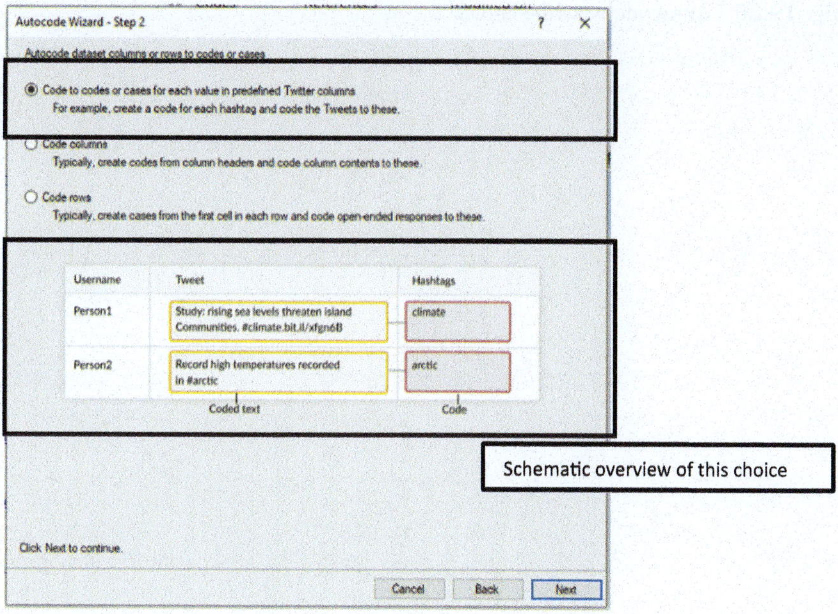

Fig. 19.19 Autocode wizard—choice 1

shown), the final choices about the name and location of codes, cases or folders must be determined.

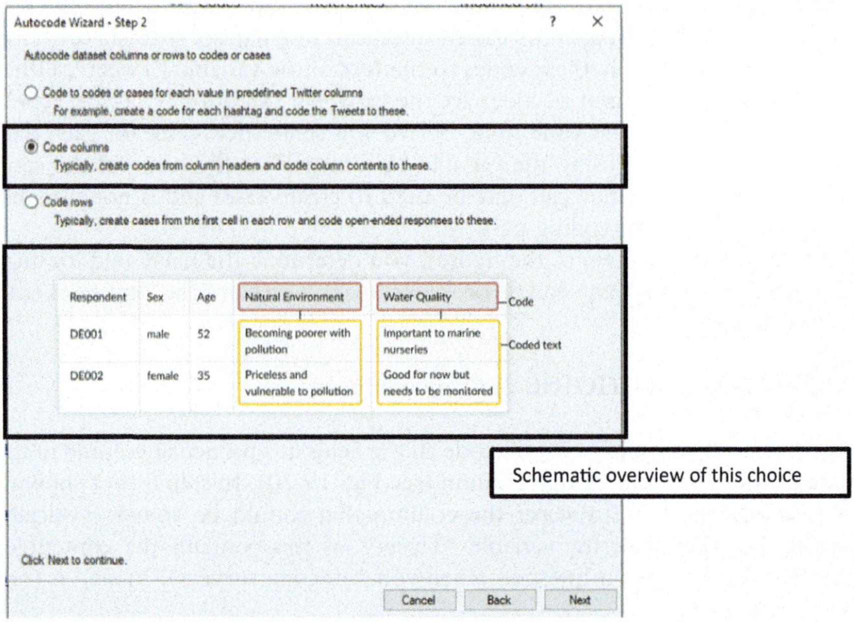

Fig. 19.20 Autocode wizard—choice 2

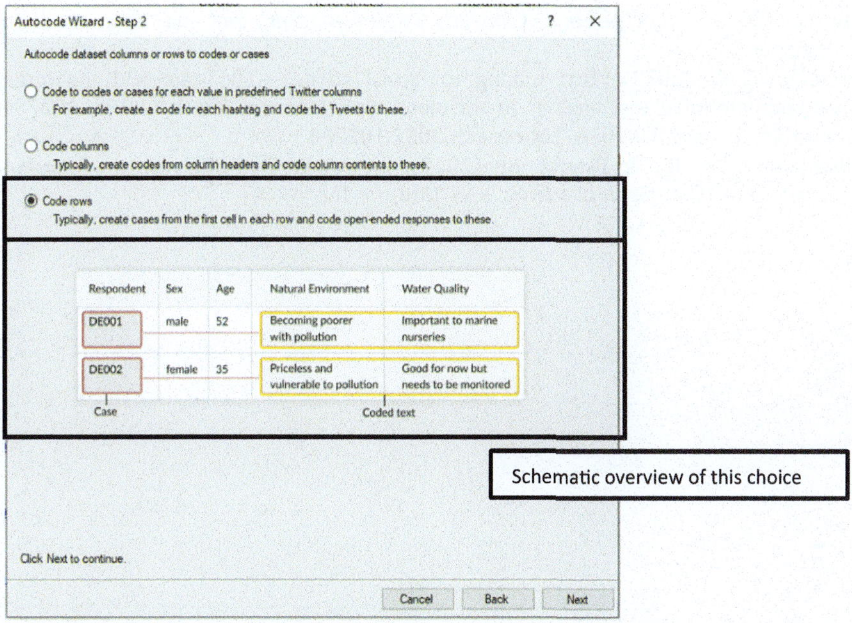

Fig. 19.21 Autocode wizard—choice 3

Autocode Wizard CHOICE 3: Code rows

The last option (see Fig. 19.21) is the opposite of the second one and codes rows. Typically, cases are created from the variable "Username" (Step 3, not shown) (or the ID variable in a survey dataset). For each user, one case is created. In the fourth step (also not shown), you determine the content that needs to be coded with these cases. Again, the most used variable here is "Tweet", as this contains the actual data of the social media (or an open question in your survey). In the final step, you again choose location and type (parent case or folder) to store the new cases.

References

Christou, P. A. (2023a). The use of artificial intelligence (AI) in qualitative research for theory development. *The Qualitative Report*. https://doi.org/10.46743/2160-3715/2023.6536

Christou, P. A. (2023b). How to use artificial intelligence (AI) as a resource, methodological and analysis tool in qualitative research? *The Qualitative Report*. https://doi.org/10.46743/2160-3715/2023.6406

Ciechanowski, L., Jemielniak, D., & Gloor, P. A. (2020). TUTORIAL: AI research without coding: the art of fighting without fighting: Data science for qualitative researchers. *Journal of Business Research*, *117*, 322–330. https://doi.org/10.1016/j.jbusres.2020.06.012

Jo, T. (2019). *Text mining. Concepts, implementation, and big data challenge.* Springer.

Macanovic, A. (2022). Text mining for social science—The state and the future of computational text analysis in sociology. *Social Science Research, 108*, 102784. https://doi.org/10.1016/j.ssresearch.2022.102784

Pinheiro, C. A. R., & Patetta, M. (2021). *Introduction to statistical and machine learning methods for data science.* SAS Institute Inc.

CHAPTER 20

Using NVivo in Team Projects

Key messages in this chapter

- Only in the server product *NVivo Collaboration Cloud*, teams can work together in one NVivo project file.
- Outside the server product, team members can distribute either password-protected projects or reports to team members.
- Outside the server product, teams can work with the standalone version of the software if they structure their collaboration to avoid data duplication or data loss.

Stand Alone or Server?

So far in this book, we have never referred to the software user. We made no assumptions about the reader of this book and their potential role as a single researcher or a team member using NVivo in the team. In this chapter, we will look at the latter situation: what if multiple researchers work together on a project, either as a group of researchers using the software together in the same project or in the role of supervisor of somebody else that is using NVivo in their project.

Working in a team with NVivo can be approached in two ways. First, NVivo is available in a server version called *NVivo Collaboration Cloud*. That means that the NVivo project of the team is installed on a server and is approachable from either a Windows or MAC computer. All changes are done locally on

D. Mortelmans, *Doing Qualitative Data Analysis with NVivo*, Springer Texts in Social Sciences, https://doi.org/10.1007/978-3-031-66014-6_20

the researcher's computer but uploaded and integrated on the server into the central project. However, the issue is that this version is not integrated into the standard version of NVivo. Each researcher needs to have a license of the program on the local computer (and work in the same version as the server version), and you need to buy an additional license on the server version to work in this integrated way. The standard add-on can be bought for a team of five researchers working simultaneously on one project, but you can extend this for bigger teams (at a cost proportional to the number of users you add). The server version is compliant with GDPR (General Data Protection Regulation) regulations so that it can be used in a European context, and it is also compliant with the HIPAA (Health Insurance Portability and Accountability Act of 1996), making it fit for the US. The server version does not require researchers to be online all the time. They can work offline on a copy of the project while not connected to the internet and upload their changes once online again.

Second, you can also work in a team when you have only bought the stand-alone version of the program. In the remainder of this chapter, we discuss working in a team with NVivo from this perspective: how do you integrate the program into your team when you only have standalone versions for each researcher? As we will explain, there is no need to buy the server version. If you prepare your project organisation in advance, teams can perfectly work with separate stand-alone versions, even across OS platforms.

Separated Team Projects

The first situation one may encounter is that one researcher is actively working in NVivo. In contrast, another researcher (e.g. the supervisor) is involved in the project but not as an NVivo user. In this situation, there is no need for the second researcher to work simultaneously on the project (see Section "Integrated Team Projects"). The active researcher needs to share information about the project with the second team member. This can be done in two ways: exchanging protected projects or exchanging reports.

Exchanging protected projects

When you want to show progress in your project with a colleague, you can send the project to that colleague and protect the content of your project with a password. Two types of passwords are possible on an NVivo project: a Read/Write password and a more restrictive Read Only password. You can enter both passwords through **File > Info > Project Properties > Passwords** (see Fig. 20.1).

You either use a Read/Write password only, or you use both the Read/ Write password and the Read Only password. When you put *both* passwords (can be the same or not), a colleague can not open your project unless they have the Read Only password. Your colleague will be able to open the project,

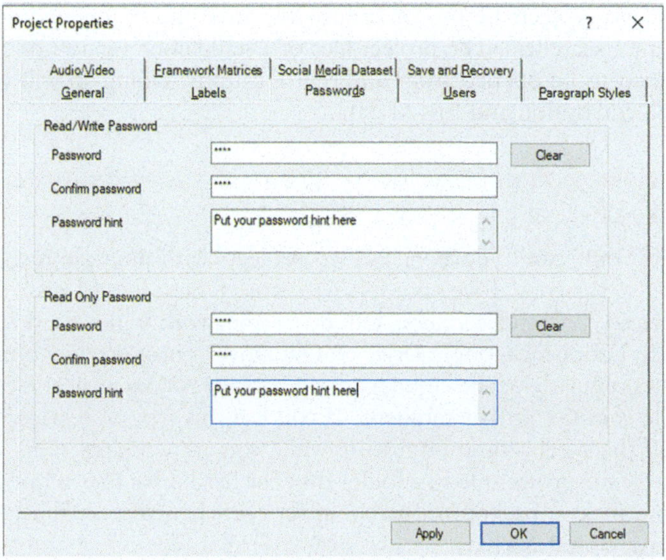

Fig. 20.1 Setting passwords in your project file

see the content and copy your project or import data from your project into another NVivo project. If your colleague wants to change your project, they will also need the Read/Write password.

When you put only the Read/Write password, a colleague will always be able to open your project and copy the project or import data from your project. When your colleague enters the Read/Write password, editing the project will become possible.

Exchanging reports

When you feel uncomfortable sending your entire project to a colleague, you might also create specific reports from your project illustrating its content and sending these reports. We have explained extensively in Section "Formatted Reports" in Chap. 14 how you make reports from your project and which reports can be made.

Integrated Team Projects

A second situation we can encounter is a research team with multiple researchers working together on a project and needing to collaborate on the same NVivo project. As stated in our introduction to this chapter, we assume these researchers do not have access to the *NVivo Collaboration Cloud*. *Still, all* have individual NVivo licenses, either on MAC or Windows (or a combination of both). A crucial caveat to start with is that under no circumstances should NVivo projects be put on a network drive and be opened from there by different researchers. The project files of a standalone version of NVivo are not designed to be opened from different locations. Doing so will very likely lead to file corruption and loss of data.

Remark

- Not only can't you open NVivo projects with multiple researchers on a network drive, but NVivo project files should also not be placed in cloud services. Often, people work with cloud services like Dropbox, Google Drive or OneDrive. These services sync their files regularly to the cloud, causing corrupt NVivo project files if the file is in use at that moment. If you use this type of service, switch off the synchronisation feature while you are working in NVivo (or copy the project file to a folder on your hard drive that is not synced and make a backup to the cloud service when you are finished and have closed NVivo).

If you do not have access to *NVivo Cloud Collaboration,* there are still ways to collaborate with the same project file safely. We present in this paragraph a way to set up such integrated teamwork, giving an example of a team of three researchers working together on one project.

This is the overview of the stepwise approach for this team:

Prerequisites before you start

- All team members have an individual NVivo license.
- One of the team members is appointed as the core NVivo-team manager (this does not need to be the team leader or supervisor of the team, but rather the team member who is most experienced with computers and NVivo). When some users work on the MAC version and other on the Windows version, it is also advisable that the team leader works on the Windows version of NVivo.

Work plan

STEP 1. The NVivo-team manager creates a core project file with all the data and folders.

STEP 2. SEPARATION PHASE: The project members receive a copy of the core project file and start working on this file (e.g. coding the file or making queries).
STEP 3. INTEGRATION PHASE: the NVivo-team manager receives all copies and integrates them into a new version of the core project file.
STEP 4. REITERATION: we start again from step 2: all project members receive a copy of the second core project file and continue their work until they reach a new integration phase. The process reiterates between the separation phases and integration phases.

STEP 1. Creation of the core project file

As three team members cannot work in one file at the same time, they must work on three copies that are compatible with each other and allow integration of coding and querying afterwards. To accomplish that goal, you start by appointing the role of *NVivo-team manager* to one of the team members. This role has nothing to do with the hierarchy in the team and does not need to be the team leader of the project or the supervisor of the project. The role of the *NVivo-team manager* goes to the team member who has the most experience with NVivo or is most experienced working with computers and software. This team member starts by creating the first version of the *core NVivo project file.* This core project file is the file that will later (in the integration phase) serve as the file to which all the work of team members will be imported. The *NVivo-team manager* is responsible for adding all data to the project and creating all folders that the team sees necessary (either in Files, Codes, or Cases).

Remark

- If not all structural elements (like folders) are known at the start, they can be added in each subsequent integration phase.

When the basic structure is defined in the project and the data is added, the first version of the core file is ready for use. The *NVivo-team manager* now creates three copies (as there are three team members, including the *NVivo-team manager*). This can be done by **File > Copy Project**. Make sure you reflect on the naming system of your files. The name of the copies should reflect (1) which researcher the file will use and (2) which version of the core file this copy originates from. As we will create multiple versions of the core file, it is important to know what version a copy is made from and to whom it

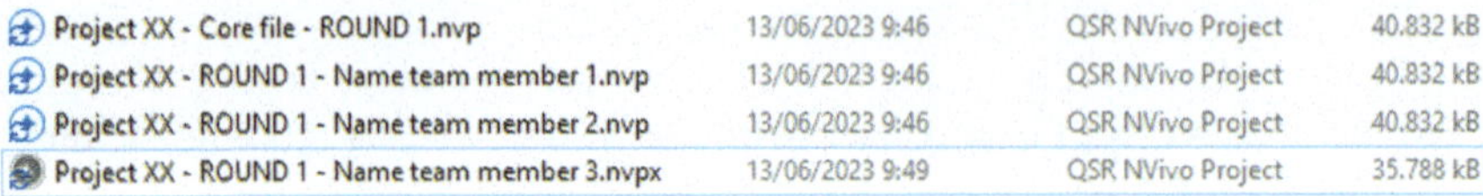

Fig. 20.2 Code file and first copies for three team members

will be sent. In Fig. 20.1, we show four files on the hard drive of the *NVivo-team manager*. The first file is the core file that contains the data and the classifications that will be used in the project. Next, the *NVivo team manager created* three copies of the core file with the names of the three team members. The file refers to round 1, the first distribution round. Also, note that team member 3 works on a MAC, and the file is in a MAC format (nvpx). This can be chosen when you **File > Copy Project** (Fig. 20.2).

STEP 2. Separation phase

In the separation phase, each team member receives a copy of the core file and starts working on that copy. The team members can create new codes and code the material. They can also make new memos or queries in the project. As a rule of thumb, they are *not* allowed to rename files, move files or change the file or case classifications. These elements all belong to the basic architecture of the project file and are the responsibility of the *NVivo team manager*. If anything needs to be changed on the data (files) or the classifications, they note their requests and hand them over to the *NVivo team manager*. In the next integration phase (see further), the *NVivo-team manager* can apply these changes if necessary.

All team members agree on the length of the separation phase. That means they agree on which moment they stop working on the separate files and send their file to the *NVivo-team manager*. When the file has been sent, the integration phase starts, and no team member can change anything in their project file any longer. All changes after sending the file to the *NVivo-team manager* will not be integrated in the next version.

STEP 3. Integration phase

The *NVivo team manager* will now integrate all working files from the team members. First, the *NVivo-team manager* makes a copy of the Core file and renames it to make it clear that Round 2 is now starting. In our example, the name of the new file would be "*Project XX—Core file—ROUND 2. nvp*".

To integrate the projects from the team members, you go to **File > Import > Project**. NVivo shows the *Import Project* window (see Fig. 20.3). The *NVivo-team manager* can choose specific items from the projects to import or can import the whole project at once. For the MAC project, NVivo will first ask to install the project converter and subsequently convert the MAC project to Windows, after which the project can be imported into the Core File.

When the import is done, the *NVivo-team manager* copies the second core file again for each team member (in the right OS format) (see Fig. 20.4).

STEP 4. Reiteration of steps 2 and 3

The *NVivo-team manager* sends the copies of Round 2 to the team members, and the team members start working on these copies. That means that we are back at step 2 (separation phase), after which the team members send their files again to the *NVivo-team manager*, and step 3 is reiterated.

This procedure seems excessively long, but it has multiple advantages. The above three team members work on their files and integrate everything. First, a core file ensures that all data and classification are unified in the merged project. As you can see in Fig. 20.3, NVivo sometimes has to deal with duplicate items. If the program cannot solve the duplication, it will simply create a copy of the item. That implies that projects that differ from each other get their data multiplied when importing. Second, the integration and redistribution at set intervals allow team members to get a glimpse of the work of their

Fig. 20.3 Import other NVivo projects

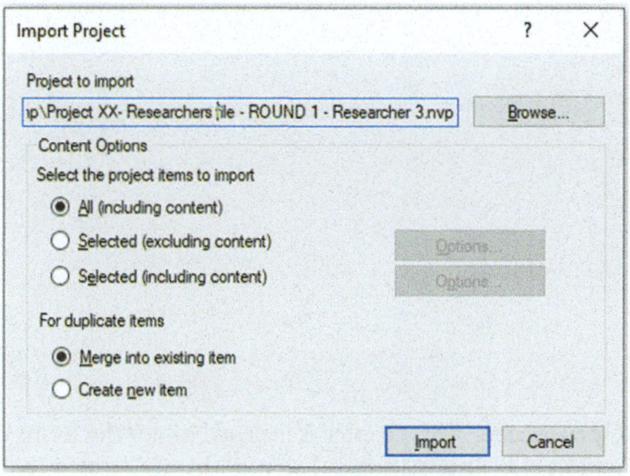

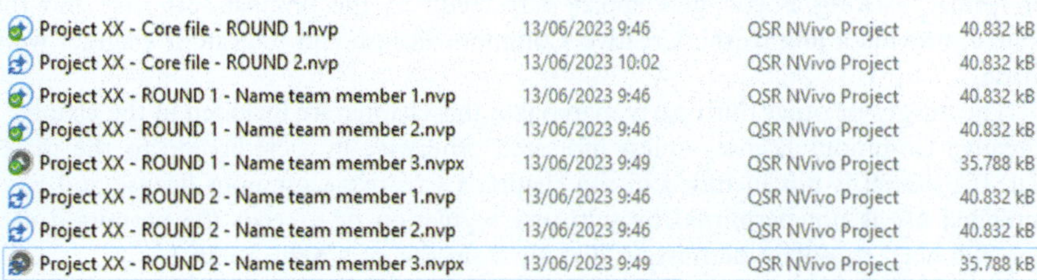

Fig. 20.4 Code file and first copies for round 2 of separating the work

colleagues. It learns how they have approached the material and which codes they have made. All team members can use the codes of their colleagues as well in subsequent rounds, leading to a tighter integration of the project. Third, the integration phase is also a natural backup of the project file on a different computer. As the files are integrated and redistributed over and over, the files are also back-upped, which ensures a minimal loss of work if anything goes wrong on the computer of any of the team members. It does not discharge the team members from their responsibility to make back-ups, but it gives an additional layer of protection to the team.

CHAPTER 21

Doing a Literature Review with NVivo

Key messages in this chapter

- With the exception of the import of reference manager databases, no specific tools for literature reviews are available in NVivo. All tools previously discussed can also be used in the context of a literature review.
- Coding literature can be done in two steps: structure coding and thematic coding. This allows a more targeted analysis afterwards.
- As literature contains more formalised language, the visualisation tools based on statistical associations (e.g. cluster analysis) work better with literature reviews.

Some Background on Literature Reviews

In this last chapter, we will not introduce any new tools. Instead, we will go over the tools we have discussed in this book and apply them in case you want to use NVivo as a program for literature reviews. As the title of this book suggests, NVivo is a program intended to help users do qualitative data analysis. That was also the program's initial purpose, and all tools were designed with this aim. As NVivo developed, the producers expanded the range of tools available to their customers. They added network and sentiment analysis. Mixed method analyses were made possible by allowing users to import

D. Mortelmans, *Doing Qualitative Data Analysis with NVivo*, Springer Texts in Social Sciences, https://doi.org/10.1007/978-3-031-66014-6_21

surveys and code open survey questions with the coding tools already available in NVivo. While tools in the program became more and more flexible and open to all types of qualitative analysis, it became clear that NVivo is also very suitable for literature review. Except for the import from Reference Managers (see Section "Some Background on Literature Reviews"), no specific tools are available in NVivo that help with literature reviews. Nevertheless, NVivo has many tools that can help users do literature reviews. Therefore, we go through the program again and show how the QDA tools can be used in writing a literature review. Before we do so, we first give an overview of the types of literature reviews that NVivo can help with.

A literature review can be done for two purposes: writing a review of the literature in and of itself and using the literature review as part(s) of a paper, book or grant proposal. The latter can be found either in the introduction of the paper, the literature section or the methodology section. In the introduction, the author presents and situates the study in the broader context of the discipline and field of study. The literature is also used to show the study's relevance and mark the current research's contribution to the literature. The literature section is self-explanatory as in this section, authors present the state-of-the-art, theoretical (what theoretical models are used in this field), and empirical (what empirical findings are there in this field of study). Lastly, the methodological section also presents the literature as the author needs to explain the methods used in the study, the ways data was gathered and justify the ways the data was analysed.

When writing a literature review, the author makes the literature review to the core of the paper. The purpose is to provide an extensive overview of literature published in a specific time frame to critically summarise and review the literature and provide some suggestions for future directions of a field. Typically, three types of literature reviews are distinguished: the narrative review, the systematic review and the meta-analysis (Efrat Efron & Ravid, 2019). A *narrative or traditional review* has a broad scope and considers various sources. A *systematic literature review* has a precise research question that is more limited in size and aims to assemble, arrange, and assess the literature on a specific topic (Paul et al., 2021) following a strict methodological protocol. The Cochrane methodology (Higgins et al., 2023) or the JBI protocol (Aromataris & Munn, 2020) are often used. A *meta-analysis* is a statistical analysis to summarise quantitative study findings showing the robustness of estimates across publications (Paul & Barari, 2022). The focus here lies on the integration of the statistical conclusions rather than presenting an overview of all insights from the literature to the reader. The difference is essential as NVivo can help with the narrative and the systematic literature review. Since the meta-analysis is a statistical analysis, NVivo cannot help create this type of literature review.

Importing Bibliographic Data from a Reference Manager

A reference manager is a software program that manages your literature in a database environment. References are broken down into the core components like the title, the author(s), the place of publication, etc. A reference manager collaborates with your word processor and allows references to be inserted in your text while you write. Using such a word processor is useful when doing a literature review in NVivo, though optional. You can do literature reviews without using them, but these programs give you a head start in NVivo when you already use them. Therefore, we will first discuss how to prepare the data from the reference manager. Next, we will show how to import the data from the reference manager into NVivo and explain how NVivo converts this data into project items.

NVivo can import data from five reference managers. One of them, Citavi, is a privileged program belonging to the same owner, Lumivero. The other four (competitors) are Mendeley, EndNote, Zotero, and RefWorks. NVivo imports elements from a reference manager and converts them into NVivo project items. This means that you need to prepare your data in your reference manager so that it can be imported optimally. We will not give an overview of the details of all five reference managers. In Table 21.1, we briefly summarise the components you can prepare in your reference manager and the project it will be imported to in NVivo.

Except for Citavi, you need to export the data to an intermediate format. NVivo cannot read the native formats of the other four reference managers. This is also visible in the menu *Import*. A separate icon, *Add from Citavi,* is shown on the main ribbon, and all five (Citavi again) reference managers are available under the icon *Bibliography*. Table 21.2 gives an overview of the different files types that NVivo needs to import data from each of these reference managers.

When you import the file, NVivo shows the *Import From* window (the example in Fig. 21.1 shows an import from Endnote).

First, you select the file to be imported. NVivo will then start to analyse the file and compare the file with the reference data already available in your project. In three sections, NVivo then gives feedback on what data will be replaced (*Already linked*), updated (*To be linked*) or newly imported (*Import new*). In each of these three sections, NVivo gives options that allow you to determine how the new or existing data needs to be treated and where it shall be imported. When you click the *Import* button, NVivo imports the data and potentially creates the four types of project items we referred to in Table 21.1: a File Classification from the reference database, PDFs from all sources found attached to the database (and on your hard drive), Externals from references not having a PDF linked (or one that is not found on the hard drive), and memos from abstracts, notes and keywords (see Fig. 21.2).

Table 21.1 Import formats from reference managers NVivo can read

	Component in the reference manager	*Project item in NVivo*
Reference database	All references in the database are broken down into their components with fields like [Title], [Author], [Publisher], etcetera	From the database, a File Classification *Reference* is created with all original fields converted to Attributes
Original source linked to a reference	Reference managers allow their users to link the data in the database to the original source (e.g. the PDF of a journal article) Remark: RefWorks's 'attachment feature' does not allow export to NVivo	NVivo looks for the original sources and, when found, imports them to the Data section as Files The PDF is also coded with the File Classification *Reference*, and the data from the source is filled in
Reference without the original source		When no link to an original source is found or when the file is not found on the hard drive, NVivo creates an External from the reference
Particular fields in the reference database	Specific fields like [Abstract], [Keywords], and [Notes]	Information from the fields [Abstract], [Keywords], and [Notes] is imported as a Memo The Memo is also coded with the File Classification Reference, and the data from the source is filled in
Subsetting the reference database	Create a group of references or a selection of references	While creating the export file for NVivo (see further), apply the selection to the file NVivo needs to import

Table 21.2 Link between a Reference Manager and NVivo project items

Reference manager	*Import format*
Citavi	– Native project file – RIS format (.ris)
Endnote	– Endnote XML format (.xml))
Mendeley	– RIS format (.txt) – RIS format (.ris)
Zotero	– RIS format (.txt) – RIS format (.ris)
RefWorks	– RIS format (.txt) – RIS format (.ris)

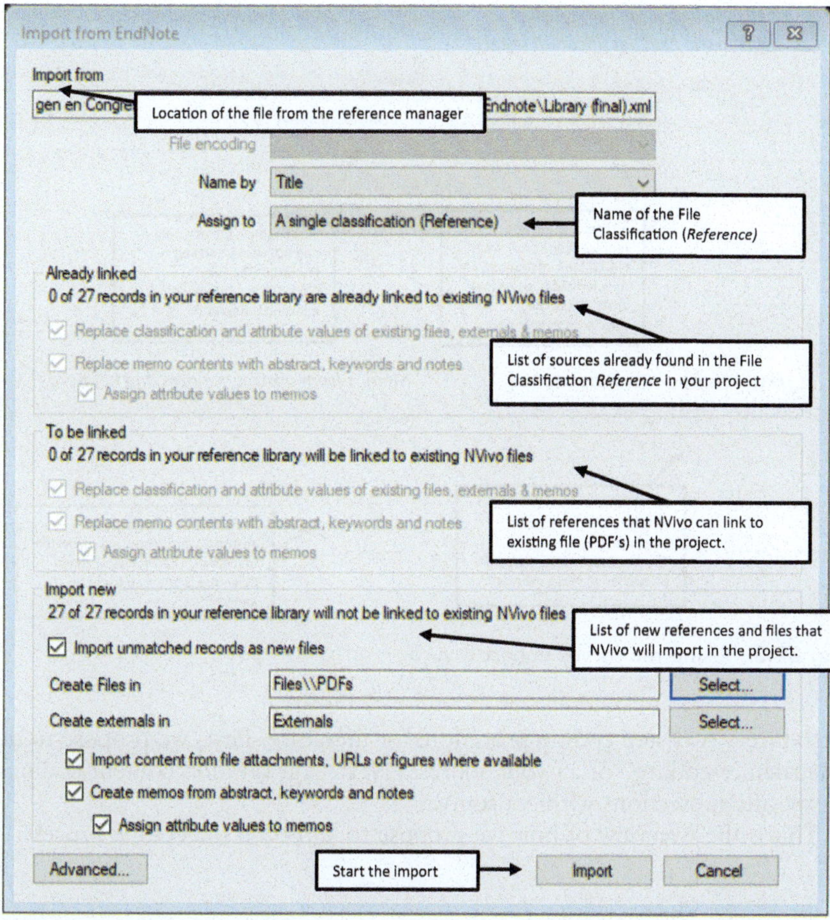

Fig. 21.1 Import bibliographic data from a reference manager (an example importing an Endnote XML file)

To use NVivo for a literature review, you need the original PDFs of your literature (books or journal articles) imported into your project. So, if the import from your reference manager did not contain all the sources you need, you must import the missing sources from your hard drive using **Import > Files**. Once imported, you can add them to the File Classification *Reference* and fill in that source's data.

Thematic Coding of Your Literature

We will not introduce not any new Coding tools in this section. We also assume you have read Chaps. 8 (Thematic Coding) and 9 (Coding with classifications) to know how coding works in NVivo. In this section, we will

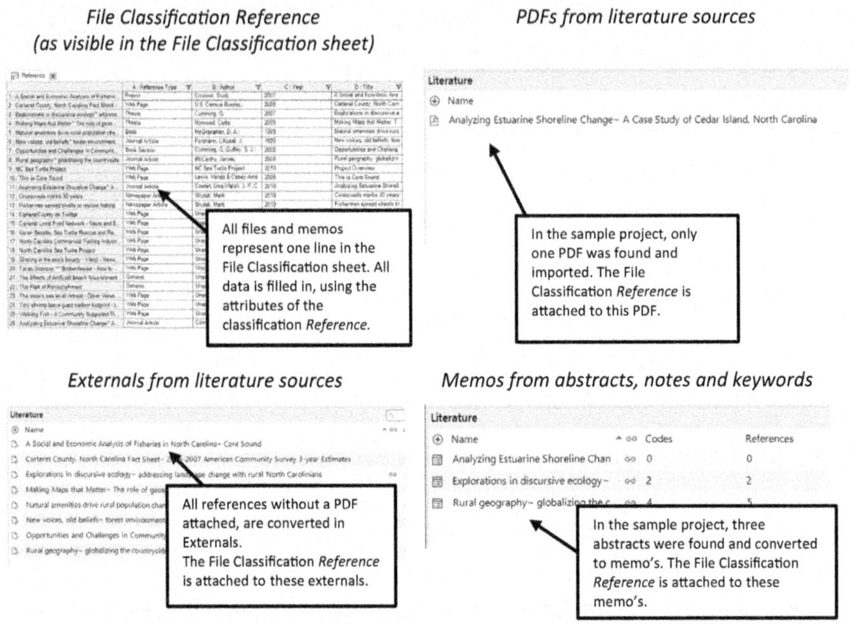

Fig. 21.2 Using the survey import wizard

introduce a two-step coding procedure for literature. First, we propose to do a "structure coding" of all your sources. Next, you can do "content-coding" on specific subsections within a source.

This is the overview of how we propose to approach the coding process:

Prerequisites before you start

- Files with literature data imported into NVivo

Work plan

STEP 1. Structure coding of all sources
STEP 2. Creating a coding query on a specific subsection of the data
STEP 3. Do manual coding on a subsection of the data in the query results window.

STEP 1. Structure coding of all sources

With "structure coding", we propose to break up the literature sources into their composing parts. Most sources, mainly journal articles, have the following sections: abstract, keywords, introduction, literature section, results section, discussion section, conclusion, and references. To identify these sections in the following steps, you select large chunks of text and code them with the section code to which they belong. This is illustrated in Fig. 21.3, where the left-hand part shows the structure codes in your codebook, and the right-hand part shows how the abstract of this article was coded with the code "Abstract". The remainder of the article will be coded with the other structure codes from the codebook.

The drag-and-drop method of coding, explained in Section "Drag-and–Drop-Coding" in Chap. 8, turns out to be very helpful in this structure coding. As the list of potential codes is limited, dragging the selections of your literature to the codebook is a very efficient way to code these sections.

STEP 2. Creating a coding query on a specific subsection of data

Once the structure of your sources is identified in Step 1, you can make a coding query to isolate a part of the literature you want to focus on. We suggest you first start with coding the abstracts of your literature. In the abstracts, the authors summarise their paper's most important elements (research questions and results). By starting to code the abstracts, you create a first version of the codebook capturing the content of your literature. In Fig. 21.4, we show how this Coding Query is defined.

STEP 3. Do manual coding on a subsection of the data in the query results window

The abstracts are shown in the results window when you run the coding query (see Fig. 21.4). This window can now be used to code the abstracts with the thematic coding techniques we presented in Chap. 8. The codebook you develop depends on the research question you posed for your literature research. In Fig. 21.5, we give an example of a codebook we developed when doing a general narrative review of the literature in our field. We focussed on definitions of concepts, theoretical models and empirical results. Use this codebook only as an inspiration, and let your research question guide you in making your codebook.

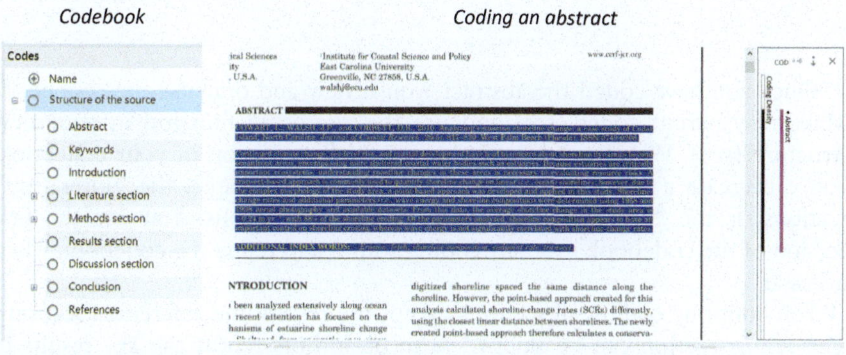

Fig. 21.3 Structure coding of literature

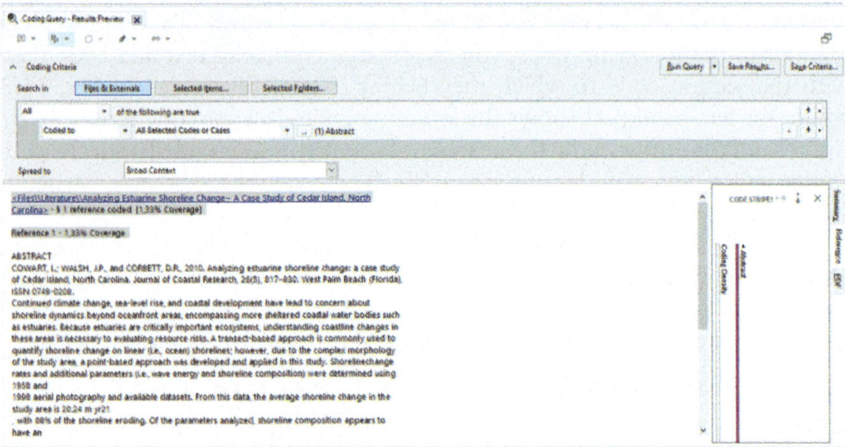

Fig. 21.4 Coding query on all abstracts

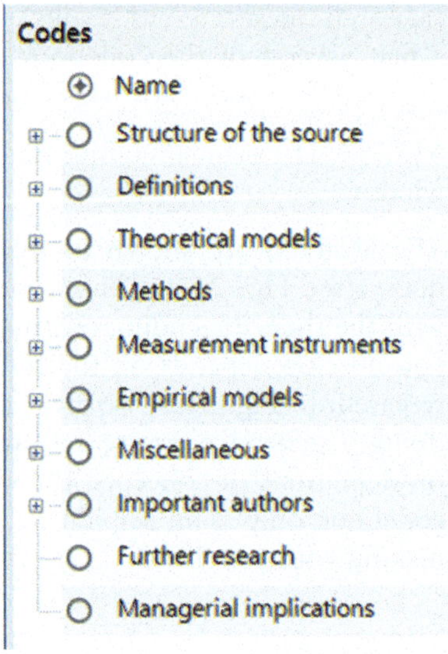

Fig. 21.5 Example of a codebook for a literature review

Once you have coded the abstract, you review and organise the codebook. Make new parent codes (see) and use the aggregate function in your new structure (see). When you have an organised first version of your codebook, you can create a new coding query showing, for example, the "Literature" sections or the "Methodology" sections of your corpus. You use the first version of the codebook in a new coding round and refine the codebook when necessary.

You continue coding all sections except the results and references sections. The *results* section can be skipped as most authors repeat the key results in their discussion and conclusion section. The references can be skipped as it is difficult to identify themes from them. The *references* section could be used later in a cluster analysis or a social network analysis (see further).

Using the Autocode Functions

An alternative to the manual, thematic coding is using the autocoding functions of NVivo. As we explained in Chap. 19, you can autocode your data using the "Identify themes" and the "Use existing coding patterns" tools. For the first tool, you let NVivo review all the literature and identify themes. For the second, you do some thematic coding first (for example, on the abstracts) and then let NVivo search the rest of the articles for similar references based on the first codebook you created manually.

You can select one article or a group of articles and start the autocoding in **Home > Autocode**. In the first step of the wizard, you choose "Identify themes" or "Use existing coding patterns" (when you did manual thematic coding, as we explained in the previous paragraph). In Fig. 21.6, we show the results of the autocoding (of themes) in the PDF in *Navigation View* > **Data > Files > Literature**.

The differences are apparent when you compare the codebook from the manual coding (see Fig. 21.5) with this codebook (Fig. 21.6). In the thematic coding, the researcher interprets what they read and can identify elements like theoretical frameworks or statistical techniques. The autocoding only identifies words that re-occur without giving context or interpretation to these terms. The interpretative work still needs to start from this codebook, and we leave it to the reader to decide what they prefer in their work. Even though the autocoding results in a codebook within minutes, the gain of time could be spurious as the reorganisation of this automated codebook can take quite a while to reach a level that approaches the thematic coding.

Summarising Your Literature

Part of writing a literature review is making a summary of the literature. You must present your readers with an overview of the state-of-the-art theoretical models and empirical findings. When you have coded your literature sources (with thematic coding or autocoding or a combination of both), you can use the Framework Matrix (see Section "Making Summaries with the Framework Matrix" in Chap. 10) to make summaries of the content of your sources.

The disadvantage of using data from a reference manager is that the database was automatically converted to a File Classification, and that there was no possibility of choosing a Case Classification for the reference data. As a Framework Matrix requires Cases, you need to convert your files to cases to use this tool (see also Section "Cases from Files" in Chap. 9). Luckily, this is made easy by NVivo. When you select all your files (CTRL + A) and then **RMC** (above the selected files) > **Create As > Create As Cases**. The result is

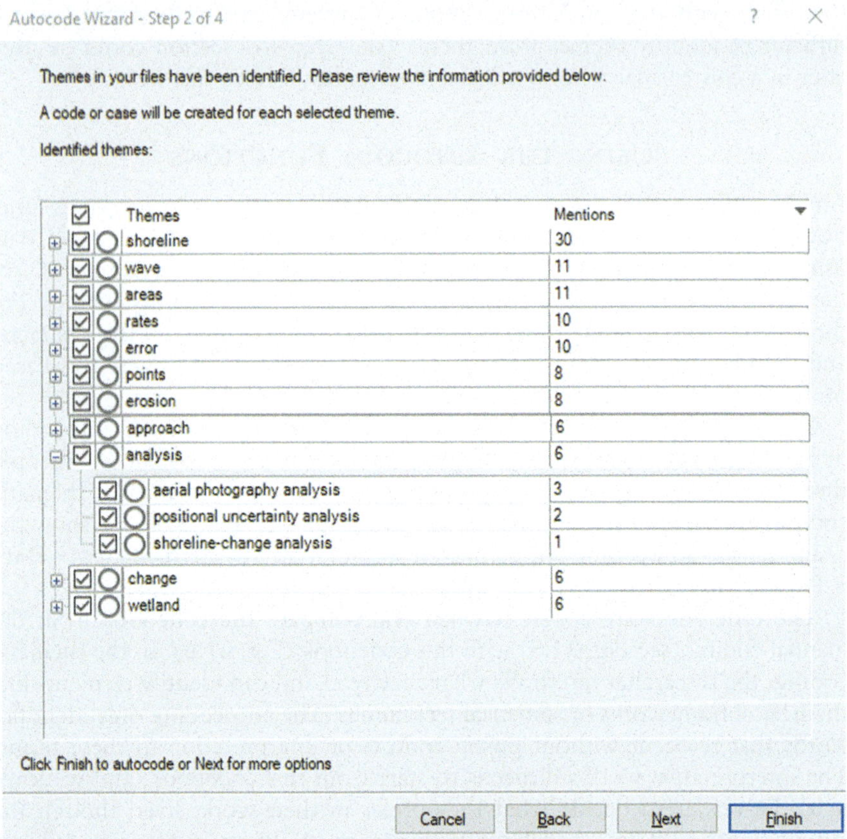

Fig. 21.6 Identified themes in a PDF of a journal article

that NVivo makes a twin case for each file and codes the files with the twin case (see Section "Making and Using Case Classifications" in Chap. 9 for more information on working with cases).

Next, you make a Framework Matrix with the cases in the rows and the codes you want to use for summaries in the Column. NVivo presents a reading window to help make the summaries, or you can use the Autosummarise function to copy the fragments to the cells where they were coded. Across one line, you get a summary of what was coded in one source. Across a Column, you can read what was coded on a certain code (e.g. a specific theory or a finding) across all sources.

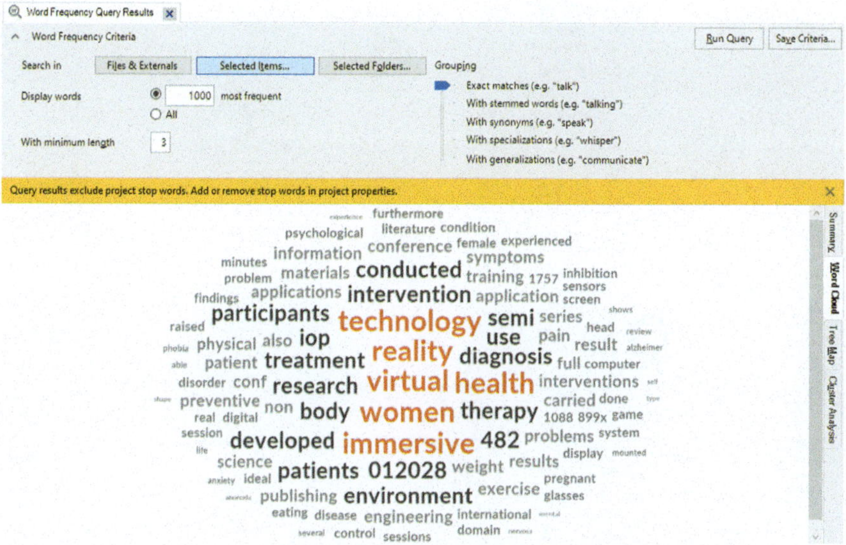

Fig. 21.7 Word Cloud (Word Frequency Query visualisation) of a journal article

Visualising Your Literature

In Chap. 10, we showed different visualisation tools in NVivo. Some were useful in interview or focus group research, and some to a lesser extent. All these tools can also be used to visualise your (coded) data in a literature review project. In this paragraph, we give a few examples of applications of these tools to literature data. We do not explain the creation of the visualisations. We only show the final result with a brief discussion. For a full explanation, we refer you to Chap. 10 on visualisations.

The Word Cloud in the Word Frequency Query

The Autocode function allows you to quickly summarise the themes in a source to a hierarchically structured codebook. But when you want a quick visualisation of the content of a source, you can use the Word Cloud visualisation from the Word Frequency Query (**Explore > Word Frequency**). It quickly shows you the most frequently occurring words in an easily accessible format (see Fig. 21.7).

The Explore Diagram

The Explore Diagram (**Explore > Diagrams > Explore Diagram**) lets you walk through your data. When you select the file of one of your sources, you get the codes attached to this PDF (see Fig. 21.8). It gives you a different visual summary of all the content in that article. But where the Word Cloud

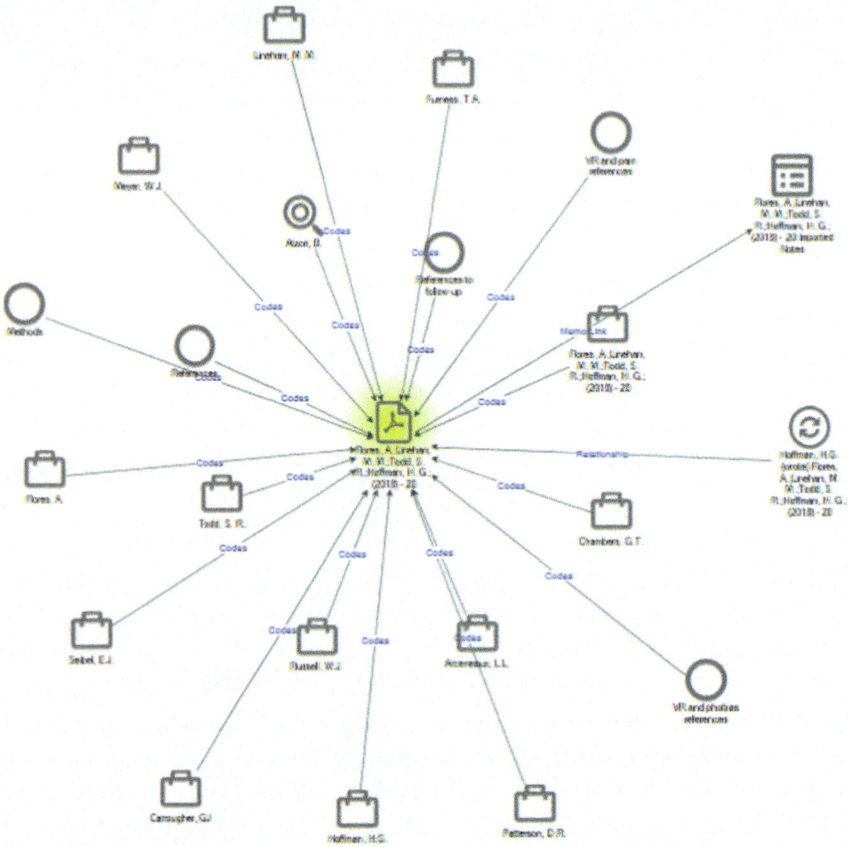

Fig. 21.8 Explore diagram of a journal article

gives you a summary of the actual words, you now get a summary based on the coding you did on this source. This might give you a different insight into the content of the source. As you can *Change Focus*, the Explore Diagram will allow you to explore themes and other sources further.

The Comparison Diagram

The Comparison Diagram (**Explore > Diagrams > Compare Files**) lets you compare two files and see what coding they have in common or not. We used this tool to compare the coding work in two review articles. We had one article published in 2009 and a second one on the same topic from 2016. The comparison diagram (see Fig. 21.9) showed us which themes disappeared from the literature (right-hand list), were stable across time (middle group of codes) or came up in the literature (left-hand side of the diagram).

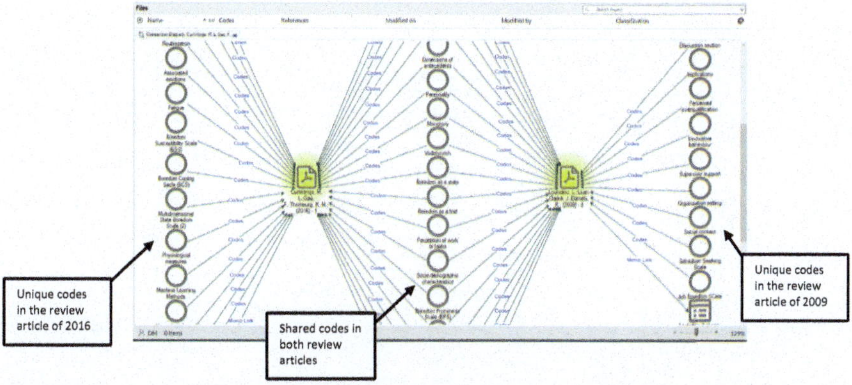

Fig. 21.9 Comparison diagram of two review articles (from 2009 and 2016)

Cluster Analysis

For interview research, performing a cluster analysis (**Explore > Diagrams > Cluster Analysis**) seems strange as this type of analysis calculates statistical associations between words in an interview. It seldom gives useful results. This is different when applying the cluster analysis to literature documents. Since academics write in a standardised, formal language, calculating associations between terms makes more sense. When two authors work with the theoretical framework of "Rational Choice Theory", they will likely use the term "rational choice" multiple times in their publications. A cluster analysis will spot this co-occurrence and place these sources close to each other in a dendrogram (this is the visual output of a cluster analysis). The result of a cluster analysis on a corpus of articles (we used the option "word similarity" when creating the diagram) is shown in Fig. 21.10. The dendrogram clearly shows groups of sources that are clustered together. It is then up to the researcher to identify the reason for their clustering.

The Hierarchy Chart

When you want a visual overview of your codebook, the hierarchy chart (**Explore > Hierarchy Chart**) is your diagram. This diagram shows the "weight" of codes regarding references attached to them using bigger and smaller boxes. Each parent code in the codebook also gets a different colour so that you can easily find a particular subsection of the codebook. Each subsection is also expandable so that you can get a hierarchy chart of all child codes within one particular parent code.

In Fig. 21.11, we show the result of a hierarchy chart on our codebook. We found that our structure codes (see Section "Thematic Coding of Your Literature") dominated the chart without giving any valuable information on the content of the analysis (right-hand chart). Therefore, we used the option

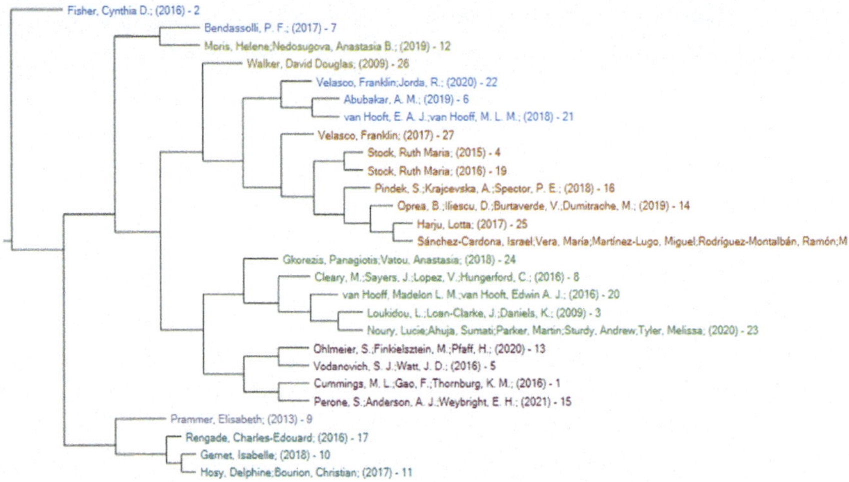

Fig. 21.10 Cluster Analysis on a corpus of literature sources

Compare Selected Items to redo the chart without the structure codes (left-hand chart). This gave a better insight into our codebook.

Querying Your Literature

In the previous paragraph, we used the Word Frequency Query to create a Word Cloud. In this paragraph, we look at two other queries that are potentially helpful in your literature research analysis. We will give examples of the Matrix Coding Query (Table 21.3) and the CrossTab Query (Table 21.4) and show their use in gaining insights into your literature.

Network Analysis to Gain Insights into Your Literature

NVivo has a Social Network Analysis tool that analyses and visualises the links between people in your study. Three types of sociograms are available: egocentric sociograms, network sociograms, and Twitter sociograms. The egocentric sociogram visualises relationships starting from one case in your project (the so-called Ego) and shows all relationships to this case. A network sociogram does not have a central case but shows how several cases are interrelated. The Twitter sociogram is intended to be used on Twitter messages and shows mentions and retweets.

di Gregorio (2021) showed how this tool can be used in a literature review, and we build upon this work to show you how social network analysis can be used in your literature study. We will also use the sample project "*Literature Review Virtual Reality and Health .nvp*" to explain this tool. This sample project can be downloaded from the start screen of NVivo (when all projects are closed) under the button *More Sample Projects*.

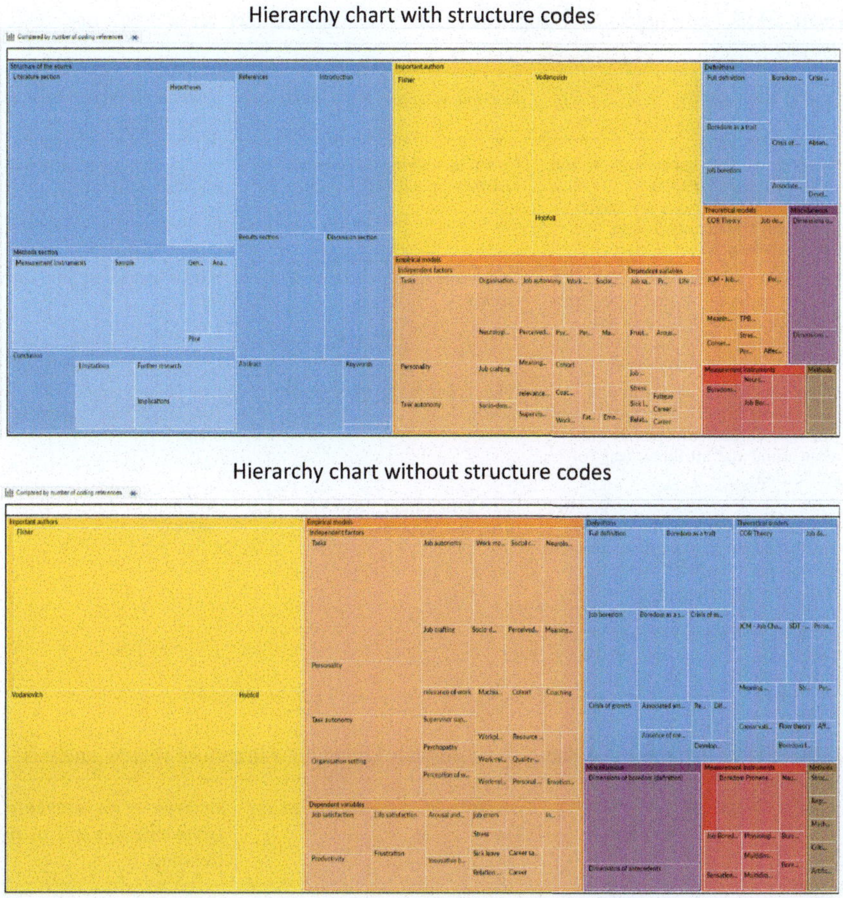

Fig. 21.11 Hierarchy chart with and without Structure coding

This is the overview of how you could perform a social network analysis in your literature review project:

Table 21.3 Examples of using the Matrix Coding Query in a literature review analysis

Aim of the example	*Example of project items in the rows*	*Example of project items in the columns*
When you coded explanatory factors and dependent variables (phenomena), you can see which have been tested together in empirical models	Codes identifying independent variables in statistical models	Codes identifying dependent variables in statistical models
When you coded on theoretical models, and you created codes for important authors, you can see which these authors are involved in which theoretical frameworks	Codes identifying theoretical models	Codes identifying important authors
When you coded on theoretical models and imported reference data, you can see how theories evolve when using the attribute Year (of publication)	Codes identifying theoretical models	File Classification "*Reference*" and Attribute "*Year*"
You can look at Journal profiles by crossing the attribute "Secondary Title" (usually, this attribute contains the Journal title) with codes or attributes containing "type of method used in a paper". It identifies which journals have a more quantitative or qualitative profile in their publication policy	File Classification "*Reference*" and Attribute "*Secondary title*"	Attribute or Code "*Type of method*."

Table 21.4 Examples of using the Crosstab Query in a literature review analysis

Aim of the example	*Examples of project items in the Rows*	*Example of the attributes (from "Reference") in the Columns*
You can look at the type of article by crossing the files of your sources in the rows with the attributes "Type of publication" (with values like "theoretical paper", "review paper", and "empirical paper") and "Type of method" (with values like "quantitative", "qualitative", and "mixed methods")	Files from your literature sources	Attribute 1: Type of publication Attribute 2: Type of method
When you coded on theoretical models, you can look at who published on these models and in what period by crossing the codes with the attributes "Author" and "Year"	Codes identifying theoretical models	Attribute 1: Author Attribute 2: Year (of publication)

Prerequisites before you start

- Files with literature data imported into NVivo

Work plan

STEP 1. Create Relationship Types aimed at literature connections
STEP 2. Create a Case Classification Authors and code your sources with it
STEP 3. Code your data with Relationships
STEP 4. Visualise with Social Network Analysis

STEP 1. Create Relationship Types aimed at literature connections

The Social Network Analysis tool is built upon the Relationships tool (see Section "Linking Data with Relationships" in Chap. 11). In a literature review, we do not encounter usual relationships like "is married to". Rather, relationships here are either between authors ("writes with") or between publications ("cites"). Therefore, we must create Relationship Types in our project to capture relationships between these types of data (authors or sources) (see also Section "Step 1. Creating Relationship Types" in Chap. 11).

In Fig. 21.12, we show all relationship types created in the sample project. It is a mixture of relationships that can be used with authors or references.

STEP 2. Create a Case Classification *Authors* and code your sources with it

As we showed earlier (see 21.2), when you use a reference manager, you have imported your reference database in NVivo as a File Classification *Reference*. This classification identifies all sources. Now, we want to analyse the level

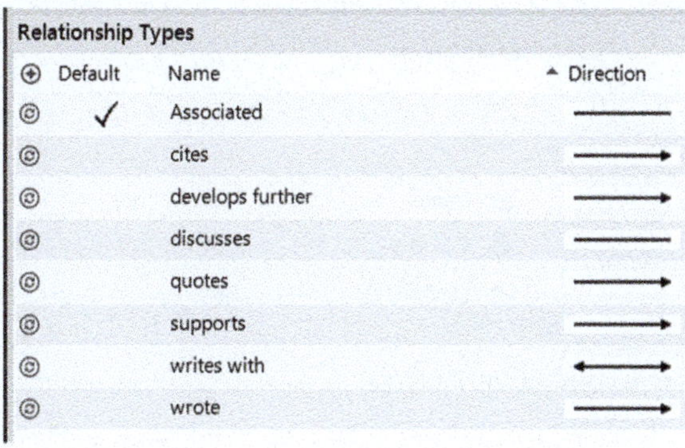

Fig. 21.12 Relationship types to use in a literature review

of authors. Therefore, we need to create a new (Case) Classification *Author* where we identify all (important) authors in our project. In Fig. 21.13, we show what this Case Classification could look like. The sample project has two attributes, *Country* and *Discipline,* but you can easily expand the number of attributes if you want more information on the authors in your project.

Next, you create a Case for each author and code the files of the publications they have written with this case. In the case, you attach the Case Classification, and you fill in the data of the two attributes (country and discipline).

The result is that you have prepared two classifications. First, you have the file classification *Reference,* and all sources are attached to this file classification with the bibliographic data from your reference manager. Second, you have a Case Classification with all individual authors as Cases and all information about the authors inside each case using the Case Classification *Authors.*

STEP 3. Code your data with Relationships

The third step involves coding your data with these relationships (see also Section "Step 2. Assigning Relationships to Project Items" in Chap. 11). The sample project gives an example of two possible ways of coding (Fig. 21.14). First, you can use the relationship type "*wrote*" to connect cases of authors to the files of their publications. For example, Hoffman "wrote" the publication Flores et al. (or at least, he was a co-author). Second, you can connect the cases of two authors with relationships. For example, Hoffman "writes with," Todd.

STEP 4. Visualise with Social Network Analysis

When all coding with relationships is done, you can create both egocentric and network sociograms. In Fig. 21.15, we illustrate the egocentric sociogram of author Hoffman. To create this sociogram, *Navigation View* > **Cases** > **Cases** and then **RMC** (*above the case of Hoffman*) > **Visualize** > **Egocentric Sociogram**.

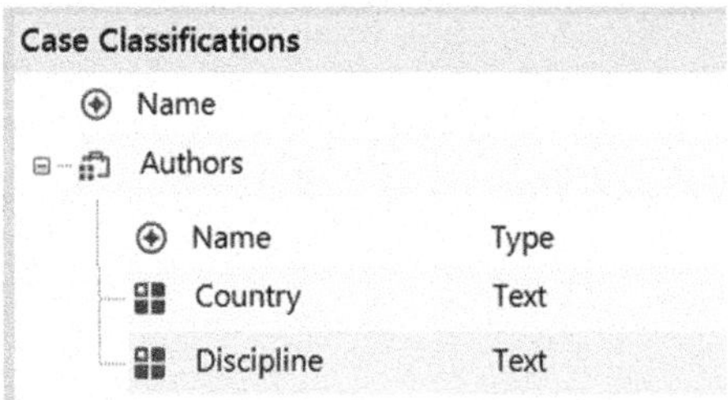

Fig. 21.13 Case Classification *Authors*

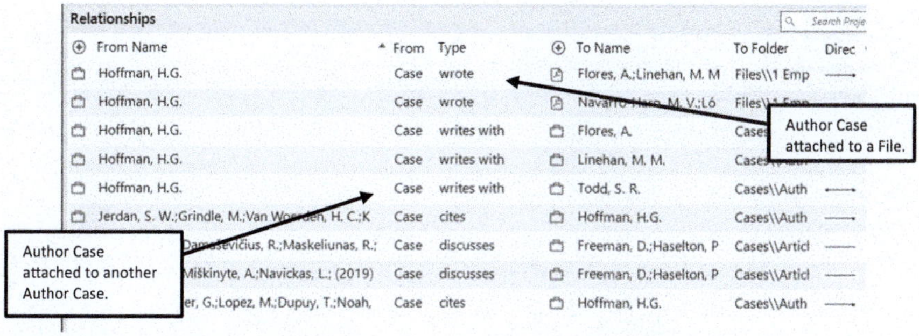

Fig. 21.14 Relationship types to use in a literature review

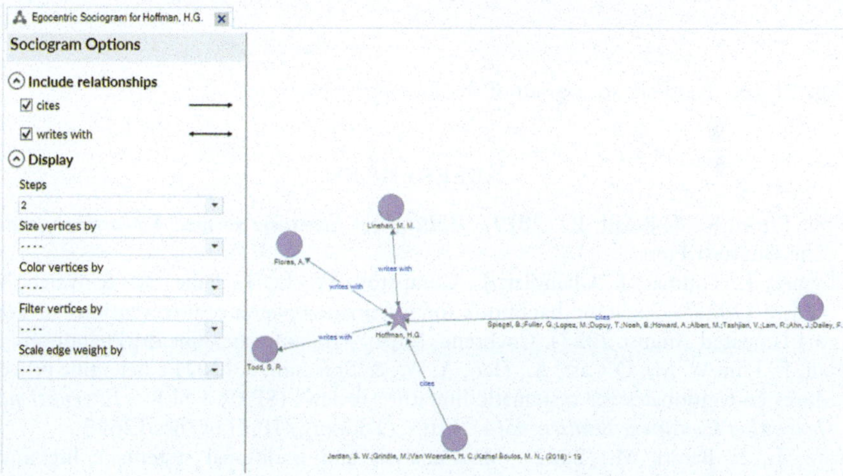

Fig. 21.15 Egocentric sociogram of one author (Case: Hoffman)

To see the interrelations between a group of authors, go to **Explore > Social Network Analysis > Network Sociogram**. In the window *Select Project Items,* you can select the cases from all authors that must be represented in your sociogram. The result is shown in Fig. 21.16.

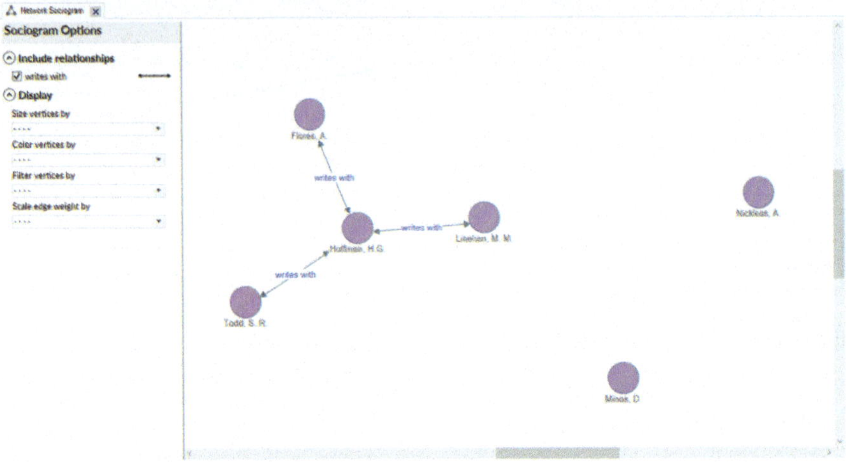

Fig. 21.16 Network sociogram of the literature review

References

Efrat Efron, S., & Ravid, R. (2019). *Writing the literature review. A practical guide*. The Guilford Press.

Higgins, J., Thomas, J., Chandler, J., Cumpston, M., Li, T., Page, M., & Welch, V. (Eds.). (2023). Cochrane handbook for systematic reviews of interventions version 6.4 (updated August 2023). Cochrane. https://training.cochrane.org/handbook

Paul, J., Lim, W. M., O'Cass, A., Hao, A. W., & Bresciani, S. (2021). Scientific procedures and rationales for systematic literature reviews (SPAR-4-SLR). *International Journal of Consumer Studies*, *45*(4). https://doi.org/10.1111/ijcs.12695

Paul, J., & Barari, M. (2022). Meta-analysis and traditional systematic literature reviews—What, why, when, where, and how? *Psychology & Marketing*, *39*(6), 1099–1115. https://doi.org/10.1002/mar.21657

Index

D. Mortelmans, *Doing Qualitative Data Analysis with NVivo*, Springer Texts in Social Sciences, https://doi.org/10.1007/978-3-031-66014-6

SPRINGER NATURE

GPSR Compliance

The European Union's (EU) General Product Safety Regulation (GPSR) is a set of rules that requires consumer products to be safe and our obligations to ensure this.

If you have any concerns about our products, you can contact us on ProductSafety@springernature.com

In case Publisher is established outside the EU, the EU authorized representative is:

Springer Nature Customer Service Center GmbH
Europaplatz 3
69115 Heidelberg, Germany

The manufacturer's authorised representative in the EU is Springer Nature Customer Service Centre GmbH, Europaplatz 3, 69115 Heidelberg, Germany. If you have any concerns regarding our products, please contact ProductSafety@springernature.com

Printed and bound by CPI Group (UK) Ltd, Croydon, CR0 4YY
09/09/2026
02213713-0002